Mathematics in
Service to the Community

Concepts and models for
service-learning in the mathematical sciences

Current printing (last digit):
10 9 8 7 6 5 4 3 2 1

Mathematics in Service to the Community

Concepts and models for service-learning in the mathematical sciences

Edited by

Charles R. Hadlock
Bentley College

Published and Distributed by
The Mathematical Association of America

The MAA Notes Series, started in 1982, addresses a broad range of topics and themes of interest to all who are involved with undergraduate mathematics. The volumes in this series are readable, informative, and useful, and help the mathematical community keep up with developments of importance to mathematics.

MAA Notes

11. Keys to Improved Instruction by Teaching Assistants and Part-Time Instructors, *Committee on Teaching Assistants and Part-Time Instructors, Bettye Anne Case,* Editor.
13. Reshaping College Mathematics, *Committee on the Undergraduate Program in Mathematics, Lynn A. Steen,* Editor.
14. Mathematical Writing, by *Donald E. Knuth, Tracy Larrabee, and Paul M. Roberts.*
16. Using Writing to Teach Mathematics, *Andrew Sterrett,* Editor.
17. Priming the Calculus Pump: Innovations and Resources, *Committee on Calculus Reform and the First Two Years,* a subcomittee of the Committee on the Undergraduate Program in Mathematics, *Thomas W. Tucker,* Editor.
18. Models for Undergraduate Research in Mathematics, *Lester Senechal,* Editor.
19. Visualization in Teaching and Learning Mathematics, *Committee on Computers in Mathematics Education, Steve Cunningham and Walter S. Zimmermann,* Editors.
20. The Laboratory Approach to Teaching Calculus, *L. Carl Leinbach et al.,* Editors.
21. Perspectives on Contemporary Statistics, *David C. Hoaglin and David S. Moore,* Editors.
22. Heeding the Call for Change: Suggestions for Curricular Action, *Lynn A. Steen,* Editor.
24. Symbolic Computation in Undergraduate Mathematics Education, *Zaven A. Karian,* Editor.
25. The Concept of Function: Aspects of Epistemology and Pedagogy, *Guershon Harel and Ed Dubinsky,* Editors.
26. Statistics for the Twenty-First Century, *Florence and Sheldon Gordon,* Editors.
27. Resources for Calculus Collection, Volume 1: Learning by Discovery: A Lab Manual for Calculus, *Anita E. Solow,* Editor.
28. Resources for Calculus Collection, Volume 2: Calculus Problems for a New Century, *Robert Fraga,* Editor.
29. Resources for Calculus Collection, Volume 3: Applications of Calculus, *Philip Straffin*, Editor.
30. Resources for Calculus Collection, Volume 4: Problems for Student Investigation, *Michael B. Jackson and John R. Ramsay*, Editors.
31. Resources for Calculus Collection, Volume 5: Readings for Calculus, *Underwood Dudley*, Editor.
32. Essays in Humanistic Mathematics, *Alvin White,* Editor.
33. Research Issues in Undergraduate Mathematics Learning: Preliminary Analyses and Results, *James J. Kaput and Ed Dubinsky,* Editors.
34. In Eves' Circles, *Joby Milo Anthony*, Editor.
35. You're the Professor, What Next? Ideas and Resources for Preparing College Teachers, *The Committee on Preparation for College Teaching, Bettye Anne Case,* Editor.
36. Preparing for a New Calculus: Conference Proceedings, *Anita E. Solow,* Editor.
37. A Practical Guide to Cooperative Learning in Collegiate Mathematics, *Nancy L. Hagelgans, Barbara E. Reynolds, SDS, Keith Schwingendorf, Draga Vidakovic, Ed Dubinsky, Mazen Shahin, G. Joseph Wimbish, Jr.*
38. Models That Work: Case Studies in Effective Undergraduate Mathematics Programs, *Alan C. Tucker,* Editor.
39. Calculus: The Dynamics of Change, *CUPM Subcommittee on Calculus Reform and the First Two Years, A. Wayne Roberts,* Editor.

MAA Service Center
P.O. Box 91112
Washington, DC 20090-1112
800-331-1MAA   FAX: 301-206-9789

Foreword

Edward Zlotkowski
Senior Faculty Fellow, Campus Compact

The Civic Engagement Movement in Higher Education

The origin of this volume lies in a project that began almost a decade ago in response to a movement that itself began to gather real momentum around 1990. The project was the American Association for Higher Education's (AAHE) series on service-learning in the academic disciplines. Early in 1995, a new faculty-based organization under the aegis of Campus Compact, the nation's only higher education association devoted primarily to civic engagement, called for the development of a series of 18 volumes on service-learning in individual academic areas. Responsibility for funding and organizing the series quickly passed to AAHE, and by 2000 the project as originally conceived had reached completion—the largest publication project in the association's history (www.AAHE.org/publications).

Drawing upon the talents of 400 contributors from every sector of higher education, the series explored both theoretical/contextual and practical issues involved in linking academically rigorous course work with projects involving the public good. English instructors explored a variety of literacy-related initiatives in community-based organizations. Biologists described not only course-based environmental research but also the creation of supplemental science resources for public schools. Accountants discussed the many ways in which participation in the Volunteer Income Tax Assistance program (VITA) could provide both a teaching resource and a public service. The medical education volume took the idea of service-learning to the professional school level. Volumes in peace studies and women's studies demonstrated the relative ease with which new interdisciplinary areas could frame syllabi organized around public issues. Since completion of this original 18-volume set, new volumes in religious studies and hospitality management have been added to the series, and a volume closely modeled on the series, *Guiding the Invisible Hand: Concepts and Models for Service-learning in Economics* (2002), has been published by the University of Michigan Press.

The movement to which these publications are a response has perhaps best been captured by the late Ernest Boyer, the individual most responsible for spurring the academy to rethink its relationship to the public good. In a 1995 address, Boyer [1] suggested that the time was now ripe for a major educational course correction—one that pointed away from an exclusive prioritizing of research and towards a deeper appreciation of the value of teaching and service as scholarship:

> ... from what I have seen, there is no question that the paradigm of faculty rewards is moving toward greater recognition of teaching. I could document that for several hours because we have the evidence in our office. I also have this sense in my bones that service is going to reemerge with greater vitality than we have seen it in the last 100 years, simply because the university must be engaged if it hopes to survive. The social imperative for service has become so urgent that the university cannot afford to ignore it. I must say that I am worried that right now the university is viewed as a private benefit, not a public good. Unless we recast the university as a publicly engaged institution I think our future is at stake. (p. 138)

It is not possible in a foreword like this even to begin to do justice to the breadth and depth of the civic engagement movement as it has developed in the academy since 1990. Hence, a few suggestive indicators will have to suffice. Despite tight institutional budgets, membership in Campus Compact has now grown to over 900 member colleges and universities, more than a quarter of the national total. During the past decade, we have also seen major engagement initiatives launched by the Council of Independent Colleges, the American Association of Community Colleges, the United Negro College Fund (i.e., private historically black colleges and universities), the National Association of State Universities and Land-Grant Colleges, and the American Association of State Colleges and Universities. Over a dozen national disciplinary associations have sponsored special projects, forums, or publications focused on engaged teaching and research.

What has made possible much of this work is generous funding from a host of public and private sources. Beginning as the Commission on National and Community Service under the first Bush administration, the Corporation on National Service (CNS) was significantly expanded under the Clinton administration and has been sustained by the second Bush administration. Its higher education Learn and Serve initiative, though modest in the context of overall CNS funding, has been especially important in helping colleges and universities develop institutional structures capable of facilitating and sustaining academic engagement efforts. Funds made available through the Community Outreach Partnership Centers (COPC) Program at the Department of Housing and Urban Development, initiated in 1994, have played a similar role. More recently, the National Science Foundation has funded a variety of projects (e.g., EPICS, SENCER) linking postsecondary science and technology education with public issues. Private foundations—among them The Pew Charitable Trusts, Ford, Kellogg, Kettering, and Lilly—have also become significantly involved.

What has been conspicuously missing from much of this activity is participation by mathematicians. Hence, even when mathematicians have shown up at cross-disciplinary, campus-based workshops on civic engagement, there has been little mathematics-specific resource material to point them to. The present volume should go a long way toward addressing this need. Although much remains to be done before academically rigorous, discipline-specific civic engagement becomes accepted as an important part of what mathematicians do and what mathematics departments offer, *Mathematics in Service to the Community* should be an important factor in encouraging and facilitating these developments and thus merits sponsorship by Campus Compact as well as the Mathematical Association of America.

The Present Volume

Readers of the following chapters, especially if they have any familiarity with civic engagement and/or service-learning, will immediately notice that the volume's editor has, as he himself admits, deliberately cast a wide net. While the framing chapters (one and six) provide a clear introduction to both the rationale for service-learning in mathematics and the concrete steps mathematicians can take to act on that rationale, the intervening chapters offer a variety of examples of not only different ways in which mathematics can be linked to projects involving the public good, but also different approaches to what is meant, at least in practice, by "service-learning."

As Hadlock and several other contributors stress, service-learning in its most effective and well-developed sense is more than another name for "real-world learning" and consists of more than applied work in the public/non-profit sector. It involves a multilayered reflection process that can substantially increase its educational value in a broad sense.

On the most basic level, reflection focuses on the mathematics to be learned and applied. One fundamental distinction between service-learning and community-service is that while the latter can be justified solely as "good deeds," service-learning can only be legitimized as a teaching-learning strategy that includes public benefits. On a second level, service-learning reflection asks the learner to become more aware of what he/she brings to the learning process: values, assumptions, biases—many of which are unexamined and potentially problematic. This second reflective level is especially important when students are involved in projects that bring them face-to-face with people that differ from them in important and thought provoking ways, such as in terms of educational level, mathematical attitudes and aptitudes, ethnic/racial traditions, or socio-economic circumstances. To leave these aspects unexplored would be to miss a vital educational opportunity, for they invariably stir up thoughts and feelings highly deserving of reflection and discussion.

A third reflective level raises analogous concerns at a societal level: over and beyond mathematical competence and personal self-awareness, how might a service-learning project lead students to consider the broad, civic dimensions of the problem they seek to address? For example, if a student successfully tutors an underserved young person and has also become much more aware of unexamined personal assumptions as a result of that work, what larger public considerations does this involvement lead to? If quantitative literacy is important for all Americans, how adequate are the resources assigned to inner-city schools, and how are decisions about these kinds of issues made? In what ways does lack of adequate quantitative literacy lead to the elimination of a long series of socio-economic options? If there is substantial room in a community for improved efficiency in operations, what does this say about the overall efficiency of government and about the use of public resources. Without asking mathematicians to remake themselves into social science teachers, we can nonetheless ask them to help their students recognize the systemic dimensions of many of our most pressing public issues. Indeed, an ability to recognize such dimensions could itself be considered a core citizenship skill.

It is in this area of more-than-just-mathematics that many of the examples offered in chapters 2–5 may still leave us with the impression of unexplored potentialities. This is not to suggest either that those examples do not succeed as applied mathematics projects in service to the community or do not in fact process student learning on a more than technical level. But it does appear that the mathematical profession may not yet have reached the stage of experience with this broader aspect of service-learning that has been achieved by a number of the other disciplines represented in the AAHE service-learning book series. Chapters 1 and 6, as well as a number of the project examples, do provide valuable advice and models for how such developments might take place. In the terminology of William Sullivan [4], this involves the process of carefully and explicitly joining "civic professionalism" and "technical professionalism" in the student learning context.

Next Steps

Limited development of the second and third levels of the reflective process is extremely common. Even in the case of service-learning projects in the social sciences, one can all too easily become so absorbed in managing the technical dimensions of the work—the dimensions, after all, that discipline-specific faculty know best and feel most comfortable with—that everything else naturally recedes into relative insignificance. And yet, if we are to re-conceptualize how the academy can make a vital contribution to 21st-century American democracy, there may be no alternative to a deliberate linkage of technical and civic competencies. Certainly the arrangement our colleges and universities typically rely on, whereby faculty deliver technical skills while personal and civic awareness is left to the catch-as-catch-can of extracurricular activities, has not served us well. Indeed, steep declines in awareness of and interest in public issues on the part of young people are what helped launch the civic engagement movement in the first place.

It is, therefore, my hope that the present book will serve as both an introduction and an encouragement to further work in civic engagement and service-learning in and through the mathematical sciences. It is, indeed, my hope that as mathematicians become more comfortable with real-world, project-based learning as a pedagogy, they will begin to feel comfortable and interested enough to seek out the kinds of resources that can explicitly embed their students' learning in a civic context and make it truly multi-faceted. As Larry Cuban [2] has pointed out, "Proponents of numeracy need to join those civic-minded and pedagogical reformers who call for tighter connections between formal schooling and life experiences" (p. 90). In other words, more successful mathematics education is ultimately a subset of a more comprehensive effort to craft an educational system that really does serve the needs of a democratic society. Mathematicians who embrace the importance of civic engagement will find they have many allies eager to assist them in their work.

I would like to conclude this foreword with a special acknowledgement of Charles Hadlock for his vision and commitment in producing this volume and of Don Albers for his leadership in making it an MAA publication.

References

1. Boyer, Ernest L. "From Scholarship Reconsidered to Scholarship Assessed." *Quest*, 48: 2 (May 1996): 129–139.

2. Cuban, Larry. "Encouraging Progressive Pedagogy." In *Mathematics and Democracy: The Case for Quantitative Literacy.* Ed. Lynn Arthur Steen. Princeton, NJ: National Council on Education in the Disciplines, 2001: 87–91.

3. McGoldrick, KimMarie & Andrea L. Ziegert, eds. *Putting the Invisible Hand to Work: Concepts and Models for Service Learning in Economics*. Ann Arbor, MI: University of Michigan Press, 2002.

4. Sullivan, William M. *Work and Integrity: The Crisis and Promise of Professionalism in America*. New York: HarperCollins, 1996.

5. Zlotkowski, Edward, gen. ed. AAHE Series on Service-Learning in the Disciplines. Washington, DC: American Association for Higher Education, 1997-2000.

Contents

Chapter 1

Introduction and Overview

Charles R. Hadlock
Bentley College

What Is Service-Learning?

Most university faculty take a broad view towards the education of undergraduates, recognizing the wide range of valuable experiences, both curricular and co-curricular, that contribute to student development during the college years. The traditional curricular structure itself places considerable emphasis on both breadth and depth, for example, although these are usually defined primarily in terms of disciplinary courses and programs. Relatively recent initiatives recognize particular transdisciplinary themes that may deserve special attention, such as communications, diversity, globalization, and service. Their sphere of relevance is not confined to one or two disciplines, nor to students in only certain major fields. But these kinds of valuable initiatives can lead a kind of orphan existence. They often rely on individual leadership in an institutional setting largely dominated by disciplinary academic departments and by the constant pressure we faculty feel to get through the syllabi for our discipline-oriented courses. So when one asks the natural questions, "What is service-learning and should I be interested in it?," it would be quite natural to do so from the standpoint of the harried character in Figure 1.1.

Figure 1.1. "Service-learning???"

If you mention the term "service-learning" to almost any mathematics faculty member, it probably conjures up the image of college students volunteering to tutor in the schools. I remember that kind of activity from college days; it was conducted as part of a community service program. Yes, it's valuable. Yes, it's a good experience for the tutors. But, no, as we will discuss shortly, without further enhancement it does not fully meet the predominant understanding of the term "service-learning" as it is widely used in the university community. The service-learning movement has been spearheaded in recent years in large part by the Campus Compact organization, a national coalition of over 900 college presidents committed to the civic purposes of higher education [1]. This has led to greater uniformity in the meaning of the term and to the development of extensive resources to support its implementation.

Service-learning may be defined quite simply as a set of activities that have two characteristics:

1. they enhance either the delivery or the impact of curricular material, usually, but not always, within the context of a specific course, and
2. they take place within a service framework where additional experience with civic engagement or social contribution will be obtained.

These two components actually fit together more naturally than might first be recognized, so let us discuss them further.

The enhancement of curricular impact can derive from several quite diverse sources, such as exposure to new techniques and ideas, motivation from seeing curricular material in action, higher student energy level due to bonding with a "client" organization and helping meet its needs, or more extensive discussion of course material due to the interactive nature of most service-learning projects. A carefully managed "reflection process," which has been well developed and studied in the service-learning field over the last decade, helps to ensure that students derive the full educational potential from such experiences [2]. Some people may think that this reflection process refers to a kind of "touchy feely" exercise that might be quite foreign to the mathematics classroom, but I prefer to think of it as the processing of a rather complex set of experiences to assure that students share and solidify their insights and thus obtain maximum lasting benefits. This has actually been one of the most important contributions of the service-learning initiative.

For some disciplines, the fact that the activity takes place within a service framework fits very specifically into the curricular application. Typical examples might be a philosophy course investigating the basis of altruism or a sociology course discussing social services or community action. Mathematics teacher education is another area where the service environment provides a clear opportunity for gaining practical experience and trying out pedagogical strategies. However, my view is that from the standpoint of mathematics instruction itself, the service focus primarily capitalizes on three important facts:

1. Many students are philosophically motivated to spend some of their time and effort in service to others, and thus they bring special energy and commitment to such activities.
2. Nonprofit and local governmental organizations often lack the resources to undertake projects or studies that they would otherwise like to have done, and thus they particularly welcome partnerships with academic organizations willing to assist them to meet

such needs. Thus community partners or "clients" with mathematically relevant needs are more readily available than one might first realize.

3. In assuming the new role of teacher or service provider, students find themselves motivated to seek a level of mastery that may well exceed their more standard profile as somewhat passive participants in a professor-dominated course.

Note that all three of these aspects can contribute to the effectiveness of mathematics learning, which is so dependent on student motivation. Without this direct benefit for mathematics instruction, it would be very hard for me to recommend service-learning to my mathematical colleagues.

Nevertheless, there are other real benefits aside from those related to mathematics itself. During our teaching careers, many of us have witnessed peaks and troughs in student activism and concern for the broader society. In my own position as a business school professor, I often worry that we can get so wrapped up in measures of financial success, both individual and corporate, that less tangible issues like community welfare and societal advancement may be given short shrift. This is one reason why we established one of the earliest service-learning programs at our own institution. We wanted our students to understand their responsibilities as members of society, and to recognize the real contributions they are capable of making even early in their careers (and even by using the material they are studying in their math classes!).

Even though mathematics faculty demonstrate in many ways their commitment to good citizenship and human values, I doubt that very many would take on the burden of incorporating service-learning in their courses simply to promote these values. Time is simply too short. However, if one does try out service-learning because of the aforementioned benefits to student learning of mathematics, I suspect that this latter aspect will emerge naturally through the discussion and reflection process. Before long, it is likely to be a theme that the instructor deliberately works in right from the start. As mathematics instructors we actually have greater power in this respect. Students expect to hear about societal problems and community needs from social science faculty, and they similarly expect to discuss these themes in connection with their readings in literature, philosophy, and the humanities. But when a mathematics professor brings up these topics, I know from personal experience that it really gets their attention.

The Status of Service-Learning in the Mathematical Sciences

In the early stages of this project, I conducted two broad surveys to learn more about the range of service-learning projects being pursued by mathematical sciences faculty. One survey was conducted through a group of service-learning directors, and the other was conducted through the MAA. From these surveys and other correspondence with mathematical colleagues, I learned about over one hundred service-learning projects in the mathematical sciences. They generally fell into three categories: mathematical modeling (at both elementary and advanced levels), statistics, and "education-related" activities (including both tutoring and other teaching-related activities). These three categories form the subjects of the three next chapters of this book.

Table 1.1 provides a brief sampling from the interesting range of service-learning activities encountered in the mathematical sciences. They arise in elementary courses such

Table 1.1. Sample mathematical sciences projects involving service-learning

Client/Organization	Project Description	Associated Course(s)
Community service organization	Design and build handicap access ramp	Algebra (slope and angles)
Homeless shelter	Tutor individual clients on "everyday" mathematics	Finite mathematics
Social service agency	Assist in the teaching of a consumer mathematics unit at an agency preparing people to own houses	Mathematics for liberal arts
Social service agency	Collect data and develop straight line approximations	College algebra
Science center	Compile visitor information and present data using tables and graphs	College algebra
Local school system	Tutoring and running special math events	Mathematics for teachers
Social service agency	Develop cost and revenue functions for various programs	Precalculus for business
Social service agency	Analyze deployment of vans and identify opportunities for improvement	Graph theory, discrete mathematics
Municipality	Design efficient snow plow routes for the city	Discrete and combinatorial mathematics
Municipality	Analyze the consequences of constructing traffic bypasses around the city	Discrete and combinatorial mathematics
Community environmental group	Analysis of traffic congestion on state highway	Statistics, operations research, senior thesis project
Social service agency	Conduct survey, compile data, perform analysis, and present results	Statistics
State environmental agency	Statistical analysis of data on fish populations	Statistics
Municipality	Quantitative analysis of risks from transportation of hazardous materials through the city	Environmental mathematics
Association of high schools	Development of questions to be used in math contests for high achieving students	Number theory, discrete mathematics
Firemen's association	Analyze implications of alternative investment strategies for an assistance fund	Mathematics of finance

as college algebra and finite mathematics as well as in more advanced courses such as mathematical modeling and discrete mathematics. Even within the single subject of statistics, the level of application is quite broad, ranging from data collection and descriptive statistics up through simple linear regression and on to some statistical testing. My own personal experience has been borne out by the many survey respondents who commented that the project framework can lead to an intense level of student motivation and to a much greater understanding of the connection between the classroom and the real world. These effects certainly promote better mathematical learning.

The Organization of This Book

From the collection of service-learning projects encountered in the course of these surveys, I chose a subset intended to give a fair representation of the range of successful projects in diverse kinds of courses at both elementary and advanced levels. These projects are discussed in individual essays in Chapters 2, 3, and 4. In some cases, the authors are using a more inclusive definition of service-learning than the rather orthodox one presented earlier. For example, several authors describe very successful, ongoing consulting programs staffed by their mathematical sciences students in which some of the consulting projects are only incidentally of a service nature, while others are not. That is, service is not a key focus, and thus the reflection process on the service aspects is likely to be informal or minimal. Such programs have been included here because of their high educational quality and because they provide exemplary frameworks for the implementation of service-learning, even if they themselves have not been fully developed along the traditional lines of this rubric. Each of these three chapters also begins with an introductory review essay by a specialist in that area, intended to compare and contrast the different kinds of projects and the various issues associated with their implementation.

While no such selection can be expected to cover the entire range of possibilities, it is my hope that these essays will provide both background information and stimulating ideas so as to assist other faculty members who may be interested in creating service-learning opportunities in their own mathematical sciences courses. One rarely transports curricular innovations in their original form; rather, one usually recreates similar innovations in the context of a different institution and along the lines of one's own interests. Thus the primary goal in selecting contributions was to stimulate thought and creativity on the part of the reader. The genuinely enthusiastic tone that permeates essentially all the contributions may provide further encouragement in this direction.

I was initially surprised by the ease of classifying the survey results into the three categories of modeling, statistics, and education. There may well be many more opportunities, some perhaps missed by the surveys, but others just so far untapped. In Chapter 5 I take the liberty of speculating on the applicability of the service-learning concept to subject matter and courses where it is not commonly found.

Implementing a service-learning course component can be a challenging task, and it may even look overwhelming if one is not aware of all the typical institutional resources that may be available. Therefore, we have included Chapter 6, which is essentially a "how to" guide to starting a service-learning initiative. This chapter is a collaborative effort between a service-learning professional and a former mathematics department chair at a large state university. Thus it represents the combined points of view of someone closely aligned with the service-learning movement and someone keenly aware of practical administrative constraints and faculty points of view.

The Educational Context for Service-Learning in Mathematics

I discussed early in this chapter the common belief of the authors of this volume that service-learning can contribute positively to mathematical learning and that it can do so in a variety of ways. Here I would like to elaborate on this theme. To begin, let us recognize the seriousness of the need: not all is well in the world of mathematics instruction. A recent report of the Committee on the Undergraduate Program in Mathematics (CUPM) cites widely held beliefs that university level mathematics departments tend to be slow to adapt to changing circumstances, especially in entry-level courses, and somewhat insular in relation to other departments and disciplines [3]. The report goes on to cite data on falling mathematics enrollments and other indicators of concern to the mathematics profession.

In fact, national studies suggest that serious issues exist in mathematics instruction at all levels, and one result has been the launching of high profile campaigns to achieve a higher level of quantitative literacy within our society [4,5]. For example, there has recently been a higher degree of formal involvement by the university mathematical community with educational strategies in the K–12 levels [6]. (In fact, several of the contributors to this volume are actively engaged in these programs.)

Discussions with colleagues support my own impression that our schools and colleges turn out millions of graduates who probably hope never to encounter mathematics again in their lives, and the majority will likely be able to fulfill that wish. The fact is that nowadays even a store checkout clerk can get on fine without being able to add. But, of course, addition is not really the issue; the role of mathematics instruction is much deeper. Can we not do better at producing students who recognize the empowering potential of mathematics and who pursue it for precisely this reason?

There is no single solution to this challenge, of course, and many creative faculty members in all kinds of institutions are actively involved in finding answers that work in their own contexts. The contributors to this volume believe that service-learning can be a valuable tool that others may also want to try. Let's look at how service-learning might complement traditional teaching in terms of meeting typical objectives for mathematics instruction, as suggested in Figure 1.2. I caution that these comments simply represent my own opinions, but I suspect that they will resonate with the observations of many colleagues. Naturally there are also considerable variations among courses, levels, and instructors.

In general, I think that traditional course instruction tends to emphasize the more concrete objectives shown higher in the figure. While I have no intention of arguing with the value of such priorities, I think that students often complete their study of mathematics without much growth in the areas that are shown nearer the bottom, and some of these latter areas may even have the most carryover value to their lives and careers. Perhaps if we found ways to target these other valuable objectives at the same time as the more concrete ones, the overall effectiveness of mathematics education would be higher than it seems to be. Service-learning may be a valuable tool in this respect.

To see what service-learning projects in the mathematical sciences might have to con-

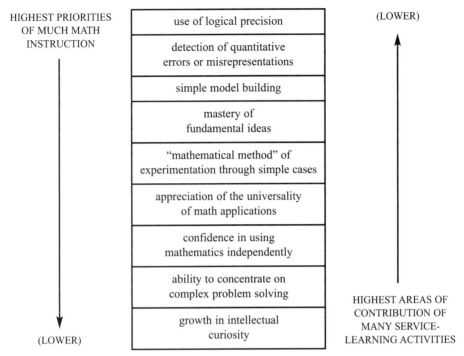

Figure 1.2. Spectrum of typical objectives of mathematics education and how these relate to both traditional instruction and service-learning activities.

tribute to this situation, let us first consider projects in the general areas of modeling and statistics. In such projects, it is typical for students to become rather deeply immersed in a client-driven problem for which they are expected to provide usable information by a fixed date. The recognition that the problem is something that they can try to solve using mathematical or statistical methods can itself be quite an eye-opening experience. For example, when I took a class on a field trip to the hazardous materials unit of a local fire department recently, the students were flabbergasted when the uniformed firefighter started talking about Gaussian plume modeling at his opening presentation. They were equally shocked when he took the group out to show them the computer where department members carry out the modeling, and the computer was inside the fire truck that goes to the scene of an accident! With an experience like this to open their minds and imaginations, the group discussion and reflection process can be an extremely effective medium for helping them recognize other surprising evidence of the universality of mathematical analysis.

The presence of an outside client, especially if it is one with whom the students develop a personal connection, is a tremendously motivating factor quite different from the typical concern about grades and pressure from a teacher. On the one hand, the students really want to deliver something useful to the client. On the other hand, they may be very concerned about the high stakes associated with standing up in front of the client organization, representing themselves and their university, and presenting their results. They may fear the embarrassment of falling flat on their faces or of disappointing a client to whom they feel really committed, while at the same time they certainly recognize that a stellar performance will be a gem for their resume and future letters of recommendation. This combination of factors can be the driving force that encourages them to experience problem solving at a new level and in a student-centered organizational structure. The fact that it is generally a team effort helps to maintain the necessary level of confidence, and if the project is successful (as well chosen projects generally are), the level of personal confidence of the individuals involved will carry through to many subsequent endeavors. The intensity of such an experience contributes significantly to the development of lasting intellectual skills.

Service-learning projects in the area of mathematics education cover quite a wide range. While tutoring is the single most common type of project reported, even this spans the range from elementary school to both college and community. In addition, there are many other kinds of creative instructional programs, both remedial and enrichment, that have been developed through a service-learning framework. They all offer some clear common benefits. For example, in preparing to teach material to others, whether via tutoring or in some other presentation format, students often find that they themselves lack an in-depth understanding of certain fundamental concepts. Thus they revisit such concepts with more motivation and curiosity and usually at a greater depth that they did when they first learned them. As further reinforcement, the actual process of explaining such concepts to others is well known to solidify one's own understanding. As a result, self-confidence in using mathematics increases, and students become more self-reliant for both learning and for catching their own errors or misunderstandings. These are the kinds of intellectual growth steps emphasized in the lower components of Figure 1.2. While these impacts should be no surprise to anyone who has taught mathematics in the classroom, it is interesting to see how consistently and enthusiastically they are also reported by the participants.

The Institutional Perspective

In considering whether to pursue service-learning initiatives, it might be valuable for a faculty member or a mathematics department to consider a broader context than the individual course or section. After all, faculty members live in at least two worlds within the university itself. One such world is the microcosm of their individual classroom, and it is the focus of much of their energy and concentration throughout the semester. The other world is the macrocosm of a complex institution, which on the one hand is driven by educational decisions of the faculty as a whole, and on the other hand is affected by the many administrative activities, such as marketing, recruiting, and fund raising, that ultimately shape the educational environment. Of course the third vital world of faculty members, that of scholarship, transcends the university boundary entirely.

From this point of view, service-learning can contribute to the overall strength of the university in number of ways, as suggested in Figure 1.3. Naturally, improved education is the central factor shown in the figure, but in this context we are not just talking about improved education in mathematics, but rather improved education of the student as a whole. There can be many educational advantages to service-learning beyond those directly related to mathematics. Examples include the following:
- development of skills in writing and speaking,
- practice in teamwork (often improved during the semester by active discussion or reflection related specifically to this issue),

- opportunities for students to exercise leadership on various portions of a project,
- exposure to the skills and protocols associated with client relations,
- the development of close relations with a faculty member or project mentor, and
- structured exposure to the theme of civic action or social contribution, which is something that might affect the student's career choices.

These advantages may not fit well within any single discipline or any set of courses, which is precisely why it is of great educational value to work them in wherever the opportunity presents itself.

The last bullet in the above list deserves special attention, for it represents a special opportunity that we as faculty have to share some of our own thoughts about the responsibilities that students will be expected to take on as they move out into the larger society. As I mentioned earlier, while students might expect to encounter issues of civic engagement or social contribution in classes in the social sciences and humanities, the impact of hearing these themes raised by a mathematics professor is probably far more powerful precisely because it is much less expected. I still remember well, for example, when a mathematics professor and mentor of mine became mayor of Urbana, Illinois. It really got my attention because it jolted my expectations. I now know well, but students may not, the deep commitment of mathematicians to political and social issues, and when we share with our students an interest in addressing such issues, as through service-learning, we can have a significant impact on their actions and their attitudes.

Another advantage that service-learning offers to the educational environment within the university is that by its very nature it encourages projects that are relevant to the real world. Students crave this kind of experience. I've had the good fortune in recent years to visit many colleges and universities to talk about real-world applied mathematics; and the

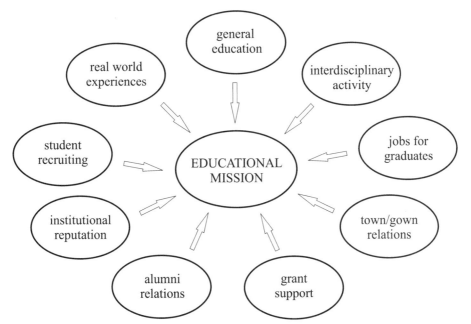

Figure 1.3. The core educational mission in relation to many contributing factors and activities that can be aided by service-learning initiatives.

level of student interest, even extending to follow-up correspondence and ongoing career advice, is testimony that this kind of experience is the as yet unfulfilled wish of many mathematics students.

Closely related to real-world issues is the growing appreciation for the importance of exposing students to interdisciplinary ventures. The organization of most universities into departments associated with individual academic disciplines is a convenient framework for managing staffing, course offerings, and budgets. However, it also has its weaknesses, one of which is the common lack of sponsors and support for interdisciplinary initiatives, such as team-taught courses or topics courses that span several disciplines. Those of us who have worked extensively outside academia have a tendency to think of almost all important applied research questions as being highly interdisciplinary, and in this framework the rigid disciplinary structure of most universities can sometimes feel quite stifling. Service-learning projects, at least those involving modeling and statistical applications, generally reach out into other disciplines, and they strengthen the university by making such interdisciplinary experiences available to its students.

Service-learning can also support other university operations, as suggested in the figure. For example, every year admissions departments undertake a vigorous and competitive campaign to attract the best class of students as freshmen for the following year. One of their most valuable marketing techniques is to highlight the special, one-of-a-kind courses offered by the university. Frequently, service-learning projects show up in this set and demonstrate the university's commitment to important educational themes, such as connections with the real world, service to the community, and interdisciplinary investigations. At the other end of the college experience, when graduates are competing for jobs or for openings in graduate and professional schools, their participation in service-learning projects that have had a real impact on the outside community can correctly demonstrate their ability to connect their academic knowledge with real-world issues. This aids the competitiveness of individual graduates and improves the reputation of the institution among employers and graduate schools. My own students often carry their service-learning reports with them to interviews.

More generally, the reputation of the university among many different constituencies has a direct impact on its ability to survive or to flourish. The institution depends in many ways on the goodwill of alumni, benefactors, government, the local community, and others, and it is not uncommon for service-learning projects to be featured at meetings with these groups. This can enhance relationships with such constituencies and occasionally leads to proposal opportunities, grants, special gifts, and other benefits.

What is the relevance of all this to the individual faculty member or department that is thinking of taking on the extra burden of developing some service-learning programs? The answer is that the university is often willing to invest concrete resources in such programs. Therefore, aside from the overall value to the institution, there are often a range of special support measures made available through the university, perhaps through a dean's office or a service-learning office, to make it easier for faculty members to get involved in this kind of activity. After all, it can take a great deal of extra time to develop a service-learning project and make it run smoothly, especially the first or second time, and it would be unrealistic for universities to expect faculty members to be able to do this without taking time from other activities for which they are also responsible. Thus it is not uncommon for special support or load adjustments to be made to encourage the further development of these kinds of activities.

Getting Started

Chapter 6 of this book provides an excellent guide for developing service-learning projects, and it also includes an annotated bibliography for the reader who is interested in accessing additional resources. My intention here is not to provide a detailed roadmap, but rather only to emphasize certain key themes to keep in mind throughout.

First and most important, there is no need to do this in a vacuum. Unless a faculty member is very well connected already with outside organizations and very clear about the specific kind of service-learning project he or she wants to develop, there is usually very valuable advice and collaborative help available from the university's service-learning or community service office. One should not underestimate the ability of such offices to adapt to a particular discipline, even if that discipline is mathematics. The staff of such offices usually spend considerable time visiting community agencies, schools, and perhaps government agencies in order to understand the scope of their operations and the areas in which they could likely use some support. Once these staff members are alerted to the kind of project that a faculty member might have in mind, they are likely to be able to suggest appropriate partners and set up initial contacts. Of course this situation can vary widely, and in some cases the service-learning office might be best for strategizing, while the faculty member may want to make direct contacts in seeking out a partner. It is certainly the case that service-learning professionals have a wide range of resources available to them, and it is likely that if similar projects have been carried out at any other institutions, they will be of great assistance in helping a faculty member find out about them.

Second, it is really important to think hard about the learning objectives associated with a potential service-learning project. Even if a faculty member is reasonably sure about the type of project he or she would like to implement, further attention to both the mathematical and nonmathematical learning objectives discussed above may help in planning the best way to implement the project, including the appropriate timing and focus of the reflection experiences that can be built into it. For example, in the hands of one faculty member a tutoring project may be simply good community service, but under the leadership of a different faculty member who has thought through the learning objectives much more thoroughly, it may be an excellent service-learning project with multidimensional educational impact for the tutors.

Third, the service value of the project needs to be real and needs to be embraced by the participating students. The first of these characteristics requires careful initial choice of the project by the professor. The second sometimes requires ongoing reinforcement during the course of the project, and it is important for the professor to monitor the attitudes of the students during the project to see whether such reinforcement needs to be orchestrated, such as by further client contact, class discussion, or even the introduction of an outside speaker to place a project in broader social context. As any teacher well knows, strong student motivation can increase accomplishment substantially.

So ... if any of these initial thoughts have raised your curiosity or captured your interest, the members of the authoring team invite you to peruse the remaining sections of this volume and to combine the ideas there with your own creative blend of interest and experience. Enjoy!

Acknowledgments

I would like to thank Edward Zlotkowski for his kind invitation to undertake this project and Don Albers for providing strong encouragement and cooperation on the part of the Mathematical Association of America. Campus Compact provided funding for production and strong support throughout. The MAA Notes Committee, chaired by Barbara Reynolds, provided encouragement and excellent suggestions and feedback. Elaine Pedreira and Beverly Ruedi were their usual cheerful and exquisitely competent selves throughout the MAA production process. Bentley College provided important support in many ways, including a reduced teaching load during part of the project. Lucy Kimball, Norm Josephy, and David Carhart were valuable collaborators as members of the editorial advisory board. Peter Noto and Liza Mattison quickly responded to messages of panic and converted my draft of Figure 1.1 into a professional drawing. My wife Joanne kindly endured my frequent periods of total obsession with the project, occasionally bizarre working hours, and other writer's quirks that she has come to know all too well by now.

Above all, I must thank the magnificent group of contributing authors, many of whom I have never met in person, who were unfailingly responsive to tight deadlines, gracious in accepting editorial comments, and, by their own deep commitment to their work as teachers in a wide range of institutions, served as an inspiration to me and as a reminder, I hope, to all of us of the nobility of the profession in which we participate.

References

1. Campus Compact home page, www.compact.org/.
2. Campus Compact. Using structured reflection to enhance learning from service, www.compact.org/disciplines/reflection/index.html.
3. Committee on the Undergraduate Program in Mathematics. Undergraduate Programs and Courses in the Mathematical Sciences: CUPM Curriculum Guide 2004, Mathematical Association of America, 2004.
4. National Research Council, Committee on Undergraduate Science Education. Transforming Undergraduate Education in Science, Mathematics, Engineering, and Technology, National Academy Press, 1999.
5. National Council on Education and the Disciplines. Mathematics and Democracy: The Case for Quantitative Literacy, ed. Lynn Arthur Steen, 2001.
6. Math and Science Partnership (MSP) program, National Science Foundation, www.ehr.nsf.gov/msp/.

About the Author

Charles R. Hadlock is Professor of Mathematical Sciences and the Trustee Professor of Technology, Policy, and Decision Making at Bentley College, where he was formerly the Dean of the Undergraduate College. He has also taught at Amherst and Bowdoin colleges and at MIT. He is the author of a prize-winning Carus monograph on Galois theory and a recent MAA text on environmental modeling. He has spent about half his career in the consulting industry, with the firm of Arthur D. Little, Inc.

Chapter 2

Service-Learning in Mathematical Modeling

2.1

Perspectives on Modeling Applications in a Service-Learning Framework

Catherine A. Roberts
College of the Holy Cross

Introduction

The purpose of this book is to explore the possibilities for service-learning within the mathematics curriculum. Service projects are an attractive way to infuse purpose and meaning into the student learning experience, and the most obvious home within the mathematics curriculum is courses that contain real-world applications. Such modeling applications can serve as the link between the classroom experience and a community-based service-learning project.

One of my favorite textbooks on mathematical modeling describes this subject as "the bridge between the study of mathematics and the applications of mathematics to various fields" [4]. A service-learning project is an ideal way to build this bridge. In this chapter, several illustrative examples of service-learning in mathematical modeling courses are presented for your consideration. Not all of these examples fit completely into the traditional concept of service-learning, yet they are featured here to help showcase the wide variety of ways in which math courses can link to the broader community, and they all suggest creative frameworks for service-learning initiatives.

It is worth emphasizing the range of possibilities for incorporating service-learning in mathematics. Projects can be integrated directly into existing courses at all levels: in section 2.6, projects for an upper-level course for majors are presented, whereas in section 2.5 and later in this section, projects for elementary level courses are described. Frameworks outside of the typical classroom can also be constructed: section 2.2 describes how teams of students work on a single project during the January term, and section 2.4 discusses a project clearinghouse that employs student workers during the summer. Section 2.3 discusses a special, project-oriented summer course. Clearly, the range of frameworks available is only limited by our imaginations.

In this section, we'll discuss the challenges, rewards, as well as the nuts-and-bolts, of developing and delivering a service-learning course. Even if you don't teach a course in mathematical modeling per se, keep in mind that most math courses contain examples that model the real world, many of which qualify as potential springboards for service-learning projects. The example that I will draw upon here is an introductory course for non-majors, Environmental Mathematics, which I have now taught twice, most recently as a service-learning course.

The hallmarks of higher education are teaching, research, and service. In the foreword to [7], R. Couto points out that service-learning and community-based research are an effective way to combine these three components of higher education. In [1], the authors note that "educators in higher education are interested in identifying increasingly better ways to achieve educational goals, including socially responsive knowledge and civic skills." Indeed, service-learning has been a recognized component of teaching in the humanities and social sciences for over two decades [6]. But what about mathematics? Does our field have a place at the service-learning table? The contributions to this chapter, indeed to this entire book, offer evidence that the answer is decidedly "yes".

The Mathematical Association of America has recently updated its Curriculum Guide for undergraduate mathematics programs [8]. Their recommendations include that departments in the mathematical sciences should "develop mathematical thinking and communications skills", "communicate the interconnections of the mathematical sciences" and "promote interdisciplinary cooperation". It's hard to imagine a more appropriate way to involve students in their communities than through a project using the analytical modeling skills that they've developed in their math course.

In this chapter, you will find five essays that detail particular initiatives that involve mathematical modeling in service-learning. They represent a variety of approaches for involving students in these endeavors. You should note that these essays are not devoted simply to sharing the end-results from successful student projects. Rather, they describe a variety of mathematical modeling applications, most within a service framework but some with a broader range of clients or objectives. In choosing to highlight these particular projects, we decided to emphasize well- developed modeling projects that underscore the kinds of programs that had great potential within a service-learning framework. The difference, of course, is that when service-type projects are chosen within these frameworks, a whole new layer of general education about civic engagement results, particularly when service is cultivated by well designed reflection exercises. Our hope is that reading through these project descriptions will support your efforts to create your own mathematical modeling service-learning initiatives.

How to Prepare and Deliver a Service-Learning Course

Service-learning courses are tailored to a particular course in most cases and to the client agencies that work with your students. The nature of this relationship can vary enormously, as can be seen in the essays that you will find in this chapter. However, there are some common considerations for anyone who is planning to develop a service-learning component that uses mathematical modeling as the link between the classroom and the community, as we discuss below.

Institutional Support

The first step is to determine whether or not you, as the teacher, are comfortable engaging in the service mission of your institution. Liberal arts colleges, research institutions, community colleges, and professional schools each view their service mission differently. But these "diverse expressions of service mission" [6] all share the goal to enhance the citizenship of their students. Students who engage in a service-learning project will inevitably grow and learn far beyond the straightforward application of their theorems and equations. Developing a successful "learning community can be far messier than a command-and-control classroom."[7] Thus, the teacher must be prepared to guide students on a learning adventure that, by design, can't be as carefully controlled as a traditionally taught math course.

The next step is to identify any institutional support that might be available on your campus. You can proceed without such support, but given the desire of many colleges to participate in service-learning, you may find that some substantial help is available. For example, could you apply for summer funding that might support the planning and design of your course? Is there a community-liaison on campus that might be able to introduce you to local civic leaders? Can you get a work-study student to help you plan the projects or coordinate the logistical details during the semester? If you bring a community leader into class as a guest speaker, can you offer an honorarium and a tray of cookies during the visit? If your students decide to create handouts, posters, or fliers, is there a funding source on campus that will pay for their duplication?

Aside from these concrete measures of institutional support, a savvy teacher will also gauge the level of departmental support for service-learning. A tenure-track colleague in another department at my institution told me that she would love to teach a service-learning course, but that when she broached the subject with her senior colleagues, it was clear that they did not support the value of such an endeavor. Even though this junior faculty member would have support from the upper administration, without departmental support it would clearly not be a prudent direction for her to pursue at this time.

Setting up the Projects

In my experience, setting up the projects takes a great deal of time. Many of my non-mathematician colleagues prefer that their students have the experience of finding a local service agency with which to partner and determining their own project. This may work with courses in sociology or history, but there are several reasons why this isn't recommended for a mathematics course. In [7], the authors note that "community-based organizations and service providers need help with their work, and they see local educational institutions as a possibility." However, the organizations themselves don't necessarily have the analytical background to clearly define a project that is doable in a semester by students at the level of your math course.

It isn't generally straightforward to conceptualize a project that involves mathematical modeling. Clients might have a sense that analytical skills would be helpful to their organization, but they generally need help defining the parameters of a project. In order for a project to be successful, the mathematical modeling can't exceed the level of mathematical expertise of your students. The students won't necessarily have the ability to make this judgment call. Secondly, the local agencies might not initially see how a mathematician can assist them, because they don't conceptualize what they need in terms of a mathematical model. On the other hand, without doing the project yourself, it's impossible to predict just how complicated the mathematical model may end up being. I've always been impressed with the Mathematical Contest in Modeling [2], where open-ended modeling problems are worked on by teams of three students over the course of a weekend. Over the years, it hasn't been unusual for a high school team to win alongside a group of under-graduate math majors with extensive math preparation. A mathematical model requires students to use whatever tools they have available, and your students can contribute pos-itively even if their mathematical backgrounds aren't terribly sophisticated. In most cases, the client will benefit from the attention your students bring to their project. For this rea-son, while it's helpful to plan the project, it's okay not to have every detail worked out ahead of time.

Identifying Partners

My entrance into community-based research happened by chance: a person from town phoned the Department looking for some help, and the secretary fortuitously forwarded the call to me. That initial conversation eventually led to a multi-year funded research project with the Grand Canyon National Park [5]. Our team, consisting of two faculty and over a dozen students (mainly undergraduates), designed a computer model to simulate white-water rafting traffic patterns on the Colorado River. This recreational use model is a planning tool for the managers in the National Park Service. The project required exten-sive field work and substantial interaction with stakeholders, even though it was essential-ly a mathematical modeling problem. The research was value-enhanced for students working in the local community: they learned how to communicate mathematics to a non-math audience, they gained an appreciation for complex political and economic issues, and they understood how our research was helping to bring stakeholders together to work towards a solution. I was hooked!

Aside from lucky phone calls, the easiest way that I've found to identify a project is to read the local newspaper. I've been particularly interested in environmental projects and my newspaper has been loaded with articles about arsenic levels in local water supplies, issues with illegal trash dumping, pesticide run-off into wetlands, etc. These stories inevitably mention a contact name that helped to point me in the right direction. I phone these people and identify myself as a teacher who is seeking a project in the local com-munity for my students.

Perhaps this is a sad comment on how mathematics is perceived in the general popu-lace, but I've found that I've had the best success when I don't mention mathematics right away. It's not always the case (and certainly wasn't with my Grand Canyon project), but I've found that some people who don't regularly use mathematics tend not to recognize its enormous potential. Consequently, they'll shut the door before we've had an opportu-nity to brainstorm potential ideas. Some people figure that if their problem doesn't already look like an algebraic equation, there's no place for mathematics. Even my faculty col-leagues who are fully engaged in service-learning from other disciplines were flummoxed to imagine how a mathematician might join in. I've found that once I can suggest possi-ble ways that mathematical modeling could assist them, the potential clients can begin to imagine the benefits to our collaboration.

Local, state, and federal government agencies are a natural for consideration as part-ners in projects for students. Nongovernmental organizations, particularly non-profit and grassroots groups, can be even better choices. Their direct need for volunteer assistance is likely to be greater and more immediate. They are also less likely to be mired in the kind of bureaucracy that can make it difficult to set up an appropriate short-term project.

In today's world of decreasing budgets, the research capacities within government agencies are often quite minimal or nonexistent. Nonprofit agencies sometimes fill in the gaps, but there is enormous need for data collection and analysis of all types. Much of this could be quite appropriate for mathematics students. The key is for students to participate in a mathematically meaningful way. Students could certainly participate in the actual data collection. However, if they can also analyze the data for emerging patterns and perhaps build a mathematical model with predictive capabilities that could assist the agency, then we can consider the project a true success.

Handling the Logistics

For my course in environmental mathematics, I contacted local non-profit agencies a full year in advance of offering the course. I wanted to develop a professional working relationship with the clients ahead of time. Depending on the project, sometimes several on-site meetings were needed to develop the project concept. It was especially important to me that these projects enhance student learning while simultaneously making a genuine contribution to the local agency.

In the end, I selected nine potential projects for the course. The 23 students formed teams that opted to work on six of the projects over the course of the semester. During the first week of class, we had a field trip to one site. Moreover, three representatives from the local nonprofit agencies were guest lecturers who visited the class to give descriptions of their agency and the needs of their projects. Students then selected projects, and all but one student received his or her first choice.

Several students worked with a local ecology museum called the Ecotarium. Their project monitored visitor use and their analysis provided input for the museum's Master Plan. Other students conducted weekly water quality monitoring tests on an urban river that the city hopes to make fishable in the near future. Other students spent weekends stenciling storm drains to advise the urban population against dumping into the local water system. Other students did projects related to asthma, mercury, and recycling in the urban environment. In each case, the service project was coupled to the course work by a final report. These reports examined the mathematics associated with their project, which in most cases correlated with the main topic of the course, that of the dispersion of pollutants. Even if a project itself does not involve mathematical analysis, I find that when students become engaged in an applied area via some kind of practical experience, they also bring a much higher level of inquiry and engagement to the related mathematical material we are discussing in the classroom. In this way, it clearly promotes the learning objectives of the course.

As the instructor, my logistical role was played almost solely prior to the start of the semester. The teaching assistant kept things running smoothly during the semester. She served as the intermediary between the client organizations and the student groups. The groups met with this teaching assistant every other week throughout the semester and only met with me twice. She provided logistical assistance and would alert me to any problems.

She also provided transportation to the sites by driving our campus van. She coordinated the lunches during our Earth Day clean-up.

At the end of the semester, I met again with each of the client partners. I provided them with a copy of the students' final project and thanked them for their participation. Conversations are presently underway to plan new projects for when the course is taught again next year.

Assessing the Student Experience

In [3], the authors cite several reasons why a service-learning project is important for student learning: it gives students a chance to learn from experience, to develop a connected view of learning, to participate in social problem solving, and to enjoy education for citizenship. How does one measure the success of such lofty goals? Assessing the extent of student learning in a service-learning project requires open-mindedness and flexibility.

In my environmental mathematics course, the service-learning project is worth 300 points (or 30% of the course grade), which are broken down between participation (50 points), an individual research paper (150 points) and a final group presentation (100 points). The final presentation included poster presentations at our local Earth Day festival, as well as at the college's research symposium. Even within these designations, however, there was a need to be flexible. For example, one project called for extensive work off campus, whereas another only required students to be off campus for one week. In each case, the students were full participants; it was just that the nature of the projects differed. Due to the variations between projects, it was important to be open to measuring the success of each project individually.

Because a service-learning project demands an extraordinary amount of student time and effort, it should count for a substantial part of the course grade. In my course, the community-based project counts for thirty percent (30%) of the grade. When students were asked at the end of the semester to comment on the relative weight of their project, about 80% agreed with its counting for thirty percent of their grade, with the remainder about equally split between assigning more and less weight.

There are many ways to assess student learning. Having various deliverables (for the agency, the broader community, or for your institution) gives students the opportunity to express their results in different ways. To assess participation, feedback was collected from the students, the sponsoring agency, and my teaching assistant. The quality of the final papers, the poster presentations, and the student feedback that I received all helped me to assess the merit of each project. Colleagues at my institution promote having students keep a reflective journal on their experience as a way to assess whether student learning has taken place.

To provide a sense of the nature of the student feedback that I solicit, here are comments from students who were asked to respond (anonymously) to the questions "Do you think this course should remain a community-based learning (CBL) course? What was the best part of the course?"

> The opportunity to give back to the Worcester community was extremely rewarding. I think it brought us closer as a class. The media attention we got was very exciting and a great way to get the issue out to the public.

The course fits in well with the "applied math" theme to the course. Going to the local elementary school and getting to show the kids the asthma peak flow meters and feeling like we really made a difference and contribution to Worcester's knowledge about asthma was the best part of the project.

I think this course was an excellent first-hand learning tool and that the students became more committed to their projects and to the class. I was able to bond with more students in this class, and the best part was being able to present outside of class to the public.

I think that the CBL component really brings math to life. The actual stenciling was a lot of fun, we were outside, enjoying ourselves while being productive in helping the community.

As a resident of Worcester County, I strongly believe that community learning is an important part of the college experience and so this course should remain a CBL. Knowing that the Ecotarium, which needs all the help it can get, will be able to use the data we collected in their planning was the best part of the project for me.

This course is all about the application of learned techniques to real-world situations, so the community based aspect is quite valuable. I had a wonderful group and I'm sure all six of us will continue to talk in the future. I enjoyed the water testing, I really felt like I was doing something.

In closing, I was pleased with the student response to the service-learning projects in the course. I was expecting to encounter more resistance to group work. One group didn't enjoy a positive dynamic, but the agency and I decided to divide their project into two separate components in order to solve the problem. Overall, remaining flexible and open-minded helped to foster success with these projects.

The Essays on Mathematical Modeling

The essays in this chapter explain how to create a viable project out of what may initially appear to be a vague idea. Faculty roles vary, although they all identified the project and established a relationship with the benefiting local agency. The essays included in this chapter will introduce you to a variety of approaches to mathematical modeling. They help illustrate the enormous potential for collaboration with outside agencies. There are many interesting aspects worth emphasizing, but my main observation is that projects can be substantial and comprehensive, or modest and straightforward. In other words, if these ideas are of interest to you, you can start by slowly growing a service-learning component into your course and, over time, you can develop it into a more sustainable program.

One approach is to have teams of students work on a single problem for an extended period of time (one or two semesters) that culminates in a formal presentation to the agency that will benefit from the analysis. This could serve as a wonderful capstone experience for math majors. At St. Olaf College (see section 2.2), a course called the Mathematics Practicum enjoys a twenty year history with over sixty projects completed. Groups of three to five students spend the January term working on a project for a local organization. The essay describing this program is an excellent overview of how to find problems, involve students, and organize meetings to make a project successful. The insightful essay from the mathematics department at Towson University (see section 2.3) describes in detail the evolution from an initial contact with an outside agency to an extensive and successful project. In this case, the scheduling process for the Baltimore City Fire

Department was examined to see if improvements could be made to reduce the enormous amount of overtime pay that was pushing the city over budget. Here, six students enrolled in an Applied Math Lab course, which resulted in a presentation to the mayor, the fire department, and other city officials officials.

Another approach is to establish a separate Modeling Research Lab that seeks out consulting contracts with local clients. In this case, students might be paid a stipend in lieu of earning course credit. Such is the case at the College of Wooster (see section 2.4), whose Applied Mathematics Research Experience is an eight-week summer program where teams of students, under the advisement of faculty, work for clients who pay a fee for services. The essay in section 2.4 describes two projects that included a strong community service component – creating an interactive web-site for Ohio farmers to obtain information about pest management and creating a web-site that helps speech and voice pathologists assist disabled clients in making decisions about which communication devices can best assist them. Such programs, where clients pay a fee for service, are best grown slowly – the authors here state that initially the project fees were quite modest as they grew their reputation. Other consulting labs for mathematical modeling exist – for example, Harvey-Mudd College and Worcester Polytechnic Institute have similar fee-for-project programs.

A truly successful service-learning project must enhance the student's sense of community and civic engagement. Whether or not the project is generating income, it can still serve as a valuable community-based learning experience for the students. There is still a service component in the sense that agencies are saving commercial consulting fees by hiring students, while naturally assuming a risk that the project won't be done as well as they might hope. In my opinion, the most effective projects place students into a situation where they are directly involved with client agencies. When students see how they can make a difference and help improve society, then their mathematical training carries more meaning. This philosophy fits very well at my institution, where our motto of "men and women for others" is brought to life by high student volunteerism and extensive cross-disciplinary community-based learning initiatives.

For the beginner, though, the best approach is probably to take an existing course and incorporate a project component into it. Even though this simplifies the implementation process, the impact on the client agency can be just as dramatic. At the University of South Florida (see section 2.5), for example, projects have been incorporated into calculus courses for engineering, life sciences and business students. The author discusses how modeling projects embedded in this most basic of mathematics courses help enhance the student experience while simultaneously benefiting the clients. For a specific service-learning project that was part of a course in discrete and combinatorial mathematics, see the essay from the University of Minnesota, Morris in section 2.6. In this project, students designed a cost-effective plan for the plowing of local city streets. The description of how this project fit into a standard course details the important aspects of designing a successful project, such as ensuring effective communication among the students, faculty mentors, and community agencies.

There are various course and non-course frameworks encountered in modeling projects, as illustrated in the essays that follow. In essentially every case, students work in teams. There is considerable room for organizational creativity—a complex project might be broken into multiple semesters, or be broken into sub-problems tackled by multiple teams. Upon completion of the project, a presentation to the benefiting agency is a customary end product, along with a corresponding report. Students are encouraged to reflect upon their experience, be it through journals or conversations with their peers and faculty mentors. Since setting up projects takes considerable time and creativity on the part of the faculty member, one must be sure to tap into the full range of institutional support mechanisms when first getting involved in these kinds of ventures. But the final benefits to the faculty, students and client agencies can be wide-ranging and very significant.

References

1. Robert G. Bringle, Mindy A. Phillips, and Michael Hudson, *The Measure of Service Learning: Research Scales to Assess Student Experiences*, American Psychological Association, Washington DC, 2004.
2. Consortium for Mathematics & Its Applications (COMAP), *Mathematical Contest in Modeling* (MCM), www.comap.org, 2003.
3. Janet Eyler and Dwight E. Giles, Jr., *Where's the Learning in Service-Learning?*, Jossey-Bass Publishers, New York, 1999.
4. Frank R. Giordano, Maurice D. Weir, and William P. Fox, *A First Course in Mathematical Modeling*, 3d edition, Brooks/Cole Thompson Learning, Pacific Grove CA, 2003.
5. C. A. Roberts, D. Stallman, and J. A. Bieri, Modeling Complex Human-Environment Interactions: The Grand Canyon River Trip Simulator, *Ecological Modeling*, Vol. 153, Issue 2, 2002, pp. 181–196.
6. Timothy K. Stanton, Dwight E. Giles, Jr., Nadinne I. Cruz, *Service-Learning: A Movement's Pioneers Reflect on Its Origins, Practice, and Future*, Jossey-Bass Publishers, New York, 1999.
7. Kerry Strand, Sam Marullo, Nick Cutforth, Randy Stoeckek, and Patrick Donohue, *Community-Based Research and Higher Education: Principles and Practices*, Jossey Bass Publishers, New York, 2003.
8. *Undergraduate Programs and Courses in the Mathematical Sciences: A CUPM* (Committee on the Undergraduate Program in Mathematics) *Curriculum Guide 2004*, Mathematical Association of America, Washington DC, www.maa.org/cupm, 2004.

About the Author

Catherine A. Roberts is the Editor of the journal *Natural Resource Modeling*, published by the Rocky Mountain Math Consortium. She is an Associate Professor of Mathematics at the College of the Holy Cross in Worcester, Massachusetts and is on the board of directors of both the Association for Women in Mathematics and the Resource Modeling Association. Her PhD in applied mathematics is from Northwestern University (1992). In addition to researching nonlinear integral equations, she spent several years working on a mathematical model that simulates white-water rafting traffic patterns on the Colorado River for use by the Grand Canyon National Park as a management tool.

2.2

Real World Consulting: The Saint Olaf Mathematics Practicum

Steve McKelvey
Saint Olaf College

Abstract. This section discusses a long running mathematics course requiring small groups of students to address problems of current interest proposed by off campus organizations and agencies, during a one month January term. There is also advice for finding projects, managing student groups, setting up final presentations at agency offices, and adapting our course design to other campuses and academic calendars.

Introduction

As the polite applause subsided, the nervous students took their seats in the conference room at the Fortune 500 company's headquarters. After a moment of thought the corporate vice president rose and, in front of several other high ranking company officials, turned to the students and said, "We just paid a consultant $50,000 for a similar study and your work is much more useful and much more impressive than what our consultants provided for us."

So ended one of the sixty projects undertaken by the mathematics practicum during the course's twenty year history at Saint Olaf College.

The mathematics practicum at Saint Olaf is an opportunity for advanced mathematics students to apply their knowledge to mathematical problems of real concern culled from sponsors in the corporate world, nonprofit organizations, and government agencies. Each January, groups of three to five students work intensively on a single applied problem of genuine interest to an off campus constituency. The course ends with a visit to the sponsor's offices, where the students make a well rehearsed, professional presentation to whomever the sponsor chooses to invite. The problems are hard; there is no textbook with answers in the back; the students get scared and the students love it.

Even at the collegiate level, a common concern of mathematics students remains, "What is this stuff good for?" For many, the inherent beauty of mathematics is motivation enough for advanced study, but even the most liberated student has some interest in learning about the application of mathematical ideas to areas outside academia. The practicum is a wonderful opportunity for students to experience, first hand, the power of mathematics.

The practicum is also a course that offers genuine benefits to the sponsors. My experience has been that many sponsors agree to participate out of a sense of civic responsibility and a broad desire to support educational activities. By the end of the course the sponsors are often greedy, in the best sense of the word, and cannot wait to see the students' results and to begin applying them to their operations.

The practicum has attracted a wide variety of sponsors, both nonprofit and corporate, over the years. Here is a representative sampling of the kinds of projects we have done for government agencies and nonprofit community organizations:

- Minnesota Department of Public Health, *Mathematical Modeling of AIDS*. From the very early days of the AIDS epidemic, this project modeled the potential future spread of the virus through Minnesota's population.
- Minnesota Pollution Control Agency, *Mapping Airborne Pollution Concentrations*. Provided a scientific basis for using a limited number of air quality monitoring devices to provide statewide air quality maps.
- Saint Paul Minnesota Blood Services, American Red Cross, *Phlebotomy Team Scheduling*.
- The Mayo Clinic, *Routine Diagnostic Test Scheduling*. Analyzed historical diagnostic test demand in an attempt to reduce patient waiting time at testing stations.
- Saint Olaf College, *Prepaid Tuition Plans*. Provided financial analysis to help the college develop various guaranteed and prepaid tuition plans.
- Saint Paul Public Schools, *Cost Effective and Delicious Menu Planning*. Incorporated student preferences, nutritional information and food costs into a newly designed menu planning system.
- Minnesota Department of Human Services, *Racial Disparities in Child Protective Services*. Helped the department fulfill a legislative mandate to explain large racial disparities in Minnesota's child protective services.

On the commercial side, we have done projects for Northwest Airlines, the Saint Paul Companies, Target Corporation, Honeywell Corporation, Intel Corporation, Cargill Corporation, Lutheran Brotherhood Insurance, and a plethora of smaller companies.

The types of projects undertaken include both statistical analyses and many types of mathematical modeling. We have produced queueing theory models, stochastic and deterministic simulations, curve fitting and optimization models. We have also encountered projects that were more theoretical in nature. Looking for efficient ways to solve a special class of systems of nonlinear equations is an example that arose in a study of aerodynamic stability and the space shuttle.

The Mathematics Practicum is an exciting opportunity that can be modified to fit a variety of academic schedules and needs. The course is very intense, both intellectually and emotionally. It is important for both teachers and students to understand this before committing to such an undertaking. While we do not formally advertise this course as a service-learning opportunity, for many of the projects that we undertake that is precisely what it is. Not only are the students using the project framework to learn about significant applications of mathematics and analytical thinking, but they are also developing an appreciation of their own capabilities to use their academic backgrounds to render significant service to an outside organization, often in the nonprofit or government sector. Furthermore, the intense full time commitment they make to this project during a one-month school term leads to extensive reflection and discussion on the process itself, on the role of the client organization, and on the utility of such contributions.

Academic Environment

Saint Olaf College is a selective private residential undergraduate liberal arts college with about 3000 students. It is located in Northfield, Minnesota, about an hour south of Minneapolis and Saint Paul, and remains actively affiliated with the Evangelical Lutheran Church in America. Most of our students are studying full time, living on or very near campus. They typically begin their studies at Saint Olaf immediately after graduating from high school. A majority of our students come from the upper Midwest and remain in the region after graduating.

Saint Olaf uses a 4-1-4 calendar, the 1 being a four week long January term during which students take a single class. We utilize this January term to offer the practicum. As you will see, the focus that this short term affords us is extremely beneficial to the course.

The practicum is team taught by two professors. Enrollment is typically capped at fifteen, with these students eventually divided into three groups that are quite autonomous during the course. As the only course being taken by our students during the January term, our students are also expected to spend a significant part of their evenings and weekends working on the project.

Finding Potential Problems and Sponsors

Finding sponsors and potential problems is the first step in the annual process of offering the practicum. While seasoned practicum professors are always on the lookout for promising sponsors and problems, the problem-recruiting process begins in earnest in early October.

Sources of leads for potential problems include the following:

- Potential sponsors finding us as a result of past experience with the course, hiring a former practicum student, or hearing about the course from past sponsors.
- Former sponsors seeking help with a new project.
- Course faculty making cold calls to interesting organizations.

Potential sponsors often have two significant concerns. First, they frequently wonder whether our call will evolve into a request for money. (It does not.) Second, people intrigued by the idea of sponsoring a practicum problem are often concerned about the time commitment we seek. We assure them that we are not asking them to teach or actively oversee any part of the course. We ask only that they agree to work with us to clearly define the project and that they make themselves available to attend the final presentation at the end of the month.

Once contact has been established, the serious work of problem definition and delineation of responsibilities begins. The first hurdle is often to identify situations within the sponsor's organization that are amenable to mathematical modeling. Many prospective sponsors do not know that they could benefit from this type of analysis. It is typically up to the faculty members to be creative in identifying opportunities for the sponsor.

After an opportunity is identified, it is extremely important for the faculty and sponsor to clearly identify the nature and scale of the proposed project. Specific attention to the nature of the final product of the course, the deliverable, is critical.

Issues of data inevitably come up during these negotiations. It is typical for practicum problems to involve some type of data from the sponsor, either for statistical analysis or to calibrate mathematical models. Practicum professors should be prepared to negotiate formal privacy agreements with practicum sponsors. It is not unusual for a sponsor to ask both professors and every student involved with a project to sign a privacy agreement. We try to make sure the agreement allows students to reveal the existence of the project and the general nature of work undertaken. This is important to students in terms of their resumes and academic records.

The second data-related issue is the timely delivery of data to the students. Unfortunately, the staff responsible for data are often extremely busy with other time-sensitive projects, and the practicum work often takes lower priority. If data delivery is postponed past the beginning of the course, students can be quickly and seriously demoralized. The faculty member must be certain that the sponsor has firmly committed to a specific time for data delivery.

In evaluating potential practicum projects, we look for problems that are nonstandard in some way, problems for which no textbook answers exist. If the entire path to solution seems clear, the problem is probably not challenging enough for a group of solid math majors with four weeks on their hands. If no first step is apparent, it is probably too difficult. If the first few steps suggest themselves, but it is not clear where they lead, the problem is probably a good one for the course, challenging enough for the time at hand yet promising enough to create reasonable hope for a solution.

Faculty Roles in Group Work

Over the years I have encountered a wide range of opinions on the faculty role within groups. There is universal agreement that practicum faculty should work to ensure healthy group dynamics and morale, and to facilitate a group's work in simple ways, like identifying vocabulary relevant to an idea developed by a group.

Opinions become more diverse on the question of how involved faculty should become in the actual mathematical or statistical analysis of the problem. It is the natural instinct of most people who have chosen mathematics as a profession to leap at the chance to solve interesting problems, and most student groups welcome the involvement of an expert in their mathematical struggles. Some believe this involvement is helpful to the students, providing them with a role model.

As you can probably guess, I find myself disagreeing. I believe that the practicum works precisely because the students know they must rely on themselves to solve the problem. It is my contention that the energy and determination practicum students commit to their group and their problem arise from the certainty that they, alone, are responsible for the group's progress. This dynamic is severely disrupted if the students believe they are being directed by professors, or that the professors know the one true answer and will eventually reveal this truth to the group.

Starting the Course

The first day of the course is a busy one for both students and faculty. By the day's end the group of fifteen students must be split into three autonomous groups of five and be given enough information on their problem to begin productive cogitation. The two faculty make short presentations on each of the three projects, each presentation lasting about fifteen minutes. For each problem, we explain what mathematics we think is going to be useful and we explain the sponsor's stake in the problem's solution. After these presentations we ask each student to rank the three problems in the order of the student's preference, giving some indication of the strength of preference.

When matching students to projects, the student preferences are taken into account. We also consider each student's academic preparation and leadership traits. We hope to populate groups with members who are motivated to work on the assigned problem and who possess talents which, taken collectively, seem likely to produce a good final product.

After the groups have been determined, each meets with the course professors to get more details on the project, to collect any data that may be provided by the sponsor, and to start organizing the group. It is expected that each group will have made some progress by the second day of the course.

Days 4 to 15

Some, but not all, practicum projects begin with a visit to the sponsor's home site. The core of the course begins after any early site visits and continues until intensive rehearsals begin for the final presentations. During this time a daily routine sets in.

Each group meets with the professors every day for a meeting scheduled to last an hour. A typical meeting begins with members of the group reporting on the previous day's homework, including news of anything momentous that may have occurred. After this quick briefing, discussions ensue regarding persistent, substantive challenges and goals for the next day's meeting.

The main role of faculty during these meetings is to monitor group dynamics and morale, and to ensure that the group is making steady progress in reasonable directions. These daily meetings can be very short if things are going well and no unexpected impediments have arisen, or they can be long and difficult if little progress is being made or personality conflicts within the group are proving difficult to resolve.

In addition to these daily meetings with faculty, each group is expected to have at least one meeting each day without the faculty. It is during these meetings that the group organizes itself, creates work assignments, resolves any other issues needing attention, and prepares for the next meeting with the faculty.

The Last Week: Presentation Rehearsals

It is impossible to overemphasize the importance of the final presentation to the working of the practicum. The prospect of being on stage in front of a group of highly interested professionals is daunting for most undergraduate students, providing a strong motivation to work hard and seriously with the objective of having something important to share with this audience. The presentation also serves to show the students that their work is impressive beyond the academy, leaving them with a strong sense of accomplishment and competence. This is an extremely valuable experience for them, especially at this time, because many are seniors and are about to head into a series of interviews for jobs or for graduate and professional schools.

The act of reflection, a key component of any service-learning experience, also plays a big part in the design of these final presentations. This is where the day-to-day homework gives way to the big picture, seeing how the entire project comes together as a united whole. The theoretical and practical come together to form a product whose value is measured by its utility to the sponsoring organization.

The rules we have adopted for our final presentation are that the talk should last slightly less than an hour, each student presents a section individually, and each section is approximately the same length. Questions from the audience are welcome at any time and the last student to speak also moderates a question-and-answer period at the end of the talk. Our goal is to make this talk a truly professional experience for both the sponsors and the students. Every aspect of the presentation looks sharp and professional, including the students themselves. Appropriate business attire is required. Although they will be present, the faculty have no formal role in the presentations.

The last week of our course is probably the most difficult for everyone. The week is devoted entirely to crafting the final presentation. We have found that a set of three private rehearsals, followed by a public full dress rehearsal and then the actual presentation at the sponsor's offices, is a schedule that seems to work well.

On rehearsal days we generally schedule meetings with each group to last two to three hours. This allows us a few minutes to discuss dangling mathematical issues, a full hour to listen to the presentation and then another hour for detailed critiques.

It is important that each rehearsal strive to be as realistic as possible, being supported with the actual props to be used at the sponsor's office. We insist that the slides be ready at the first rehearsal, understanding that subsequent editing is likely. The first rehearsal is often an eye opener for the students as they realize how difficult it is to create a truly professional presentation. Videotaping the rehearsals and turning over the tape to the students can be a very effective motivational device.

After three private rehearsals, the presentations are pretty slick and, for the most part, finalized. The public dress rehearsals are given the day before the actual presentation to the sponsor. The students are encouraged to invite their friends. Faculty not directly involved in the course are also invited and often attend. Students from the other practicum groups, if their own presentations are under control, often attend, curious to see what the others have been up to for the month. Every aspect of the final presentation is faithfully undertaken during the dress rehearsal, including the Q&A session.

Rehearsal week is extremely difficult and draining for both students and faculty. The students typically rehearse and rewrite their sections several times each day in addition to their single performance in front of the faculty. For the faculty, rehearsal days are exhausting, consisting of three sessions of two to three hours each. Each rehearsal requires the complete focus and attention of the faculty in order to keep good notes on the successes and weaknesses of each speaker. The evaluation sessions are also very intense, as the faculty try to craft constructive criticism without diminishing confidence and pride among the students.

The Final Presentation

After all the work leading up to it, the final presentation is a relief. The students are excited and confident. The professors cannot wait to get home and get some sleep.

It is important to maintain the student-focused spirit of the practicum through the final presentation. One way to support this is for the faculty to physically locate themselves as far from the presentation as possible. Most of our presentations take place in a rectangular conference room at the sponsor's office with the students speaking at one end of the room and the sponsor's personnel sitting at a long table. In such an environment, the faculty will sit along the back wall, physically separating themselves from the sponsor's staff and the students. Our role during the presentation is to smile supportively, exuding confidence in the speakers should any glance our way.

The presentations always go well, and the sponsors are universally impressed by the quality and quantity of the work done by the students over a period of four weeks. The Q&A sessions slowly dissolve into praise for the students. Leaving the sponsor's site, the students are usually aglow with pride and released nervous energy.

Evaluation and Grading

The group nature of the practicum makes the evaluation of individual students extremely difficult. Most of the group work happens outside the view of faculty. We see each group for an hour a day and our attention is inevitably drawn to the more outspoken members. What we see or hear during these meetings is not necessarily representative of the work being done behind the scenes.

Similarly, during presentation rehearsals students are often presenting work they did not themselves create. This exercise is a good thing, but can tend to obscure who deserves credit for a given mathematical insight. In addition, much of the work of the group is logistical rather than mathematical; keeping track of references, writing software, designing slides, etc. This work is also valuable and needs to be recognized.

We grade the practicum by beginning with the idea that everyone in the group deserves credit for group accomplishments. So, at the first approximation, we expect to give everyone in the group the same grade. Minor deviations from this expectation can occur based on our observations during our daily group meetings, performance during presentation rehearsals and group member evaluations. At the end of the course we ask each group member to evaluate the performance of his or her colleagues, describing roles, reliability, work ethic, etc.

Finding Faculty to Teach It

The practicum is nothing like a traditional mathematics course. Faculty who are comfortable with the more traditional mathematics classroom, where the teacher has a pretty good idea of what is going to happen next and can steer things in a predetermined direction, are sometimes intimidated by the thought of running the practicum. For some the barrier is their technical background. They worry that they may lack the necessary skills in the areas of mathematics, computing, and statistics that are likely to come up in a practicum environment. Others are concerned about their ability to find problems in government, industry or the nonprofit sectors, where they may have few, if any, personal contacts. Lastly, some find it almost impossible to avoid working on the problems themselves.

Except for the last, these problems loom larger in the imagination of uninitiated potential practicum faculty than in reality. Teaching an unstructured course is actually quite delightful. As the teacher learns to trust his or her instincts it becomes a creative exercise on its own, with all the satisfaction that comes from any beautiful and challenging creation.

Those worried about their technical background can also rest easy. A fact appreciated by all mathematical modelers is that most modeling situations do not require research level mathematics. Part of the beauty of the practicum is the discovery that lower level mathematics, calculus, simple ordinary differential equations, linear algebra, etc., is incredibly powerful and robust in its applications. Any mathematics Ph.D. who has been teaching a collection of rigorous courses at the undergraduate level either knows or can learn the mathematics likely to be relevant to any practicum problem. After all, it is the students who are meant to become true experts on the problem, not us.

While personal networks can make identifying potential practicum sponsors much easier, these networks are by no means necessary. Most organizations are philosophically disposed to helping a good school with its educational mission. These organizations are also interested in meeting good students close to graduation. Involvement in the practicum can be a great recruiting device for the sponsor. Anyone can make the cold calls necessary to land good practicum problems. It is made easier when the caller has confidence that the potential sponsor is being offered a quality experience. The practicum is a quality experience for everyone involved.

Adaptations to Other Environments

The course I have described is taught annually at Saint Olaf College and depends on the existence of our January term and the ability of the college to assign two professors to the course. Our proximity to the Minneapolis and St. Paul metropolitan area gives us access to a large number of progressive companies, nonprofit organizations and government agencies that might be interested in working with us. What about setting up a practicum-like course in different academic environments? One of the terrific benefits of our practicum model is that it can be extended to a wide variety of situations.

At Saint Olaf we have adapted the practicum model to a second January term mathematics course, an off campus course entitled Mathematical Modeling at the Biosphere 2 Center, taught just outside Tucson, Arizona. Unfortunately, Columbia University severed its relationship with the Biosphere 2 Center in December, 2003, ending this course's run after only two offerings. Still, it was exceptionally successful during its brief existence and shows how the practicum model can be adapted to different environments.

The Biosphere course was designed for third year mathematics majors. All the key components of the practicum were present. The students undertook several independent group projects supporting ongoing environmental science research at the Biosphere 2

Center. Groups met with the lone Saint Olaf professor at least once each day to ensure progress on their projects, and the course ended with public presentations offered to interested scientists, faculty and tourists at the Biosphere. This class also included a more traditional daily lecture on mathematical modeling, covering a standard curriculum as well as mathematical topics that arose from the students' group work.

A similar course could be set up anywhere with the magic combination of appropriate classrooms, computing facilities, student residential space and a source of mathematical problems and willing collaborators.

Summer school seems to offer another opportunity for a practicum-like experience. It would seem critical that students enrolling in a summer practicum have very clear expectations of time commitments and out-of-class availability. Summer jobs and recreational expectations have the potential of creating havoc and ill will within groups.

I am not sure how a practicum-like course would work if embedded in a more traditional academic semester or quarter. An important part of our offering is the energy and focus resulting from the practicum being the sole academic demand placed upon our students during the January term. It is easy to schedule group meetings and long rehearsals. This flexibility would be seriously reduced if students were fully engaged in several courses and the usual number of extracurricular activities. It might be possible to set up a course similar to the practicum as one of a student's several courses, but the experience would be substantially different from our practicum. This is certainly true for institutions with a large part-time or nonresident student population.

Summary

The mathematics practicum at Saint Olaf is an intense, one month long, group experience in which students are given the opportunity to tackle a significant mathematical problem of practical interest to an organization while under faculty supervision. Solution of the problem and communication of the solution to nontechnical persons are both given serious emphasis. The course faculty are managers, not doers.

Students experience the power of the mathematics they have spent years learning and the relevance of mathematics to organizations and entities outside academia. Students also gain respect for the importance and complications of the missions of these organizations, be they for profit, nonprofit or governmental.

Faculty get the chance to expand personal networks into the broader community. The chaotic nature of the practicum can offer a nice change of pace for instructors. The course also boosts the visibility of the educational institution to an important local constituency. Finally, and to me, most importantly, the intensity of the practicum experience creates an unusually tight personal bond between the teachers and the students, a bond that can last well past graduation.

About the Author

Steve McKelvey is an Associate Professor of Mathematics and Associate Dean of Students at Saint Olaf College in Northfield, MN. He has been involved with the practicum since his arrival at Saint Olaf in 1985.

2.3

The Baltimore City Fire Department Staffing Problem

Andrew Engel, Coy L. May, and Mike O'Leary
Department of Mathematics, Towson University

Abstract. The Towson University Applied Mathematics Laboratory formed a team of six undergraduate students, led by two faculty advisors, in order to study staffing methods of the Baltimore City Fire Department, with a view towards minimizing overtime labor costs.

Introduction

The Applied Mathematics Laboratory of Towson University undertakes undergraduate research projects for local companies and government agencies. In Summer 2002, the Baltimore City Fire Department asked the Applied Mathematics Laboratory to analyze their scheduling process. That year the fire department was two million dollars over its budget in overtime, and they wanted to see if improvements could be made. A team of six undergraduates led by two faculty members studied the problem for two semesters. The project culminated when the student research team presented their results to Baltimore Mayor Martin O'Malley, his staff, and fire department officials at the CitiStat briefing [1] at City Hall in May 2003 (Photo 2.3.1 & Photo 2.3.2). The students' analysis suggested that the city could save as much as $250,000 per year by adjusting their staffing. This result received media attention, including a piece on a local television newscast [2] and an article from the Associated Press [3].

This was a mutually beneficial project. The fire department received a detailed staffing analysis and software tools that will allow them to study future hypothetical scenarios. Our students received a valuable introduction into the use of mathematics to solve real problems and into mathematical research in general. They gained some interesting insights into a local governmental agency, a labor union, and even Baltimore City politics. Finally, the project strengthened the ties between Towson University and Baltimore City government.

We begin with a description of Towson University and its Applied Mathematics Laboratory. Next we give a brief overview of the staffing problem. We include a formulation of the mathematical model, discuss the techniques used and give the results. We also describe the project, including how it was obtained and administered. Finally, we discuss our general experience with projects such as this.

Background

Towson University is a public university in the state of Maryland. We are a comprehensive Master's granting institution located just outside of Baltimore City. With 13,000 undergraduate and 3,000 graduate students, Towson is the second largest university in the state. Our entering undergraduate students have typical SAT scores between 1020 and 1160 with a high school GPA of 3.45.

Our mathematics department has undergraduate programs in pure mathematics, applied mathematics, mathematics education, and actuarial science and risk management. We have two Master's degree programs, one in applied and industrial mathematics and one in mathematics education. Currently we have about 120 undergraduate majors and 40 graduate students.

The Applied Mathematics Laboratory looks for local companies and government agencies with mathematical research questions at the advanced undergraduate level. We then choose a problem whose analysis, solution, and exposition requires the substantial involvement of a team of students and faculty members for an academic year. Next we form such a team of four to six talented undergraduate students, led by a faculty director from the mathematics department and a co-director with particular interest in the problem posed. At the end of each semester, a written report and an oral presentation are given to the sponsor.

Photo 2.3.1. Student team member Michael Machovec briefs Mayor O'Malley and his staff.

Photo 2.3.2. Student team member Marco Radzinschi responds to a question fromMayor O'Malley.

Student participation in Applied Mathematics Laboratory projects is by invitation only. Student participants register for courses called Applied Mathematics Laboratory I and II and receive 3 hours of course credit per semester. The faculty receive credit for teaching the courses as well as some release time.

The Applied Mathematics Laboratory was founded in 1980, and its genesis was supported by a grant from the National Science Foundation (#8160689). Since then the Applied Mathematics Laboratory has worked on fourteen different projects. Project sponsors have included the Baltimore City Fire Department, the State of Maryland Comptroller's Office, Science Applications International Corporation (SAIC), and Becton Dickinson.

One of the primary duties of the director of the Applied Mathematics Laboratory is to find sponsoring organizations with potential projects. One source of contacts is our Mathematics Associates Advisory Board, a group of people from local companies and government agencies who advise us on their mathematical needs. Another effective technique has been direct mailings to local companies and government agencies; our project with the Baltimore City Fire Department was found in this manner.

The Staffing Problem

The Baltimore City Fire Department sought help in analyzing its scheduling process to see if there might be ways to reduce overtime expenditures. The department's overtime expenses had been exceeding the budgeted amount by approximately two million dollars per year. A principal challenge was the complex set of constraints on personnel scheduling.

The primary division in the fire department is fire suppression, which includes emergency medical services. For reasons of public safety, all fire suppression positions must

Table 2.3.1. Required staffing for the Baltimore City Fire Department

	Required Staffing	Officers	Pump Operators	Vehicle Drivers	Firefighters	Other
Division 1						
Division Chief	1					1
Div. Ch. Aide	1					1
Battalion Chiefs	3					3
Engines	76	19	19		38	
Trucks	36	9		18	9	
Air Cascade	1			1		
Medics	20					20
Rescue 1	5	1		1	3	
Hazmat 1	1			1		
EMS Officers	2	2				
Division 2						
Division Chief	1					1
Div. Ch. Aide	1					1
Battalion Chiefs	3					3
Engines	68	17	17		34	
Trucks	40	10		17	13	
Air Cascade	1			1		
Medics	24					24
FireBoat 1	5	1			2	2
FRB 1	2				1	1
EMS Officers	2					2
Totals	**293**	**59**	**36**	**39**	**100**	**59**

be staffed at all times. In the fire department each time a position cannot be filled, an off-duty employee must be called back to fill this position. Since these employees are working overtime, they are paid at one and one-half times their normal rate. This practice leads to large overtime expenses in the department.

The fire suppression division is separated into four shifts. Each shift works two 10-hour day shifts, followed by two 14-hour night shifts and then four days off before resuming the cycle. Each shift requires 293 employees to fill all of the necessary positions (see Table 2.3.1). To allow for absences, approximately 372 employees staff each shift. Vacation days are allotted on a rotating basis; 1/6 of each shift is eligible to take a vacation day. On an average shift, 15% of the total shift is on vacation. An additional 11% of the staffed personnel are absent for other reasons. This means for a normal shift, on aver-

Table 2.3.2. Sample callback data

	Aug 28 Day	Aug 28 Night	Aug 29 Day	Aug 29 Night	Aug 30 Day	Aug 30 Night	Aug 31 Day	Aug 31 Night
Battalion Chief	1					2		1
Captain			1	2		3	1	2
Lieutenant	5	5	3	6	3	2	4	3
Pump Operator	3	5	5	1	2	4	1	3
Vehicle Driver	8	3	6	1	2	5	2	3
Paramedic	2	2	3	2		2		3
Firefighter/ Paramedic		1	4	3	1		3	5
Firefighter	14	9	7	11	5	11	3	13
Firefighter/ Paramedic Apprentice		3	3	3	1	6		4
Other		1		1				

age only 275 of the 372 staffed employees are available to fill the required positions. This leads to an average of 18 callbacks per shift (see Table 2.3.2 for a sample of callback data).

Rather than call back an exact replacement, the fire department allows its employees to "act out of rank." For instance, a firefighter may act as a lieutenant, pump operator, or vehicle driver. One way to fill a captain's vacancy is to call back a firefighter, with a lieutenant acting as the captain and the firefighter acting as the lieutenant. In addition, the department attempts to even out the callbacks so each employee has the same likelihood of being called back. The mathematical model of the staffing problem must consider both the callback procedure and the acting out of rank possibilities.

In addition to the problem of the fire suppression division, the fire department asked for and received an examination of their communications division staffing. The communications division is responsible for handling all incoming 911 calls and dispatching the appropriate fire department resource to handle the call. This staffing problem is quite similar to the one for fire suppression, however, and we do not treat it in any detail here.

Mathematical Model and Results

The students in the Applied Mathematics Laboratory broke the project into three primary pieces. First, they analyzed the absence data from the fire department to determine what patterns were present. Next, they constructed a nonlinear program to determine a candidate optimal solution. Finally, they built a simulation of the fire department's daily staffing which was used to determine the expected costs of the staffing level found by the optimization program.

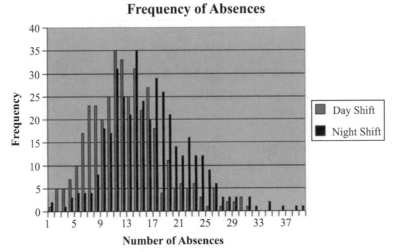

Figure 2.3.1. Histogram of non-vacation absences.

As the students began analyzing the absence data from the fire department, they originally tried to fit a binomial distribution to the absence data. This was not successful, since the vacation absences are nearly constant at 15% of the total scheduled. A hypothesis test was then employed to verify that vacation absences were constant at this percentage. Non-vacation absences appeared to follow a binomial distribution, although the parameters were different for the day shift and the night shift (see Figure 2.3.1). This was verified by a chi-square test.

To align ranks and duties in the fire suppression division, five job categories were created (see Table 2.3.3). Using these job categories and the absence statistics, the students were able to formulate the nonlinear program. The objective function of the nonlinear program is

$$\sum_{i=1}^{5} c_i(x_i + 1.5y_i) + \sum_{i=1}^{5}\sum_{j=1}^{5} r_{ij}a_{ij},$$

with c_i the salary cost of position i, x_i the number of position i to staff, y_i the number of position i to callback, a_{ij} the number in position i acting in position j, and r_{ij} the marginal cost of a person in position i acting in position j. The first constraint is that the fire depart-

Table 2.3.3. Job categorization by rank.

Job Category	Ranks	
Officer	Captain	Lieutenant
Pump Operator	Pump Operator	
Driver	Vehicle Driver	
Firefighter	Firefighter Firefighter Paramedic	Firefighter Paramedic Apprentice Paramedic
Specialized	Battalion Chief	Other

ment must staff at least the required minimum at each position, so that

$$x_i \geq m_i, \quad i \in \{1,2,3,4,5\},$$

where m_i is the minimum number needed in job i. In addition, the total number of employees actually working that job must be greater than the minimum required for that job; thus

$$y_i + \sum_{j=1}^{5} a_{ji} \geq m_i, \quad i \in \{1,2,3,4,5\}.$$

Next, the total number of employees in a given category in attendance must be equal to the number of employees of that category working a job, so that

$$(1 - p_{non})(1 - p_{vac})x_i - \sum_{j=1}^{5} a_{ij} = 0, \quad i \in \{1,2,3,4,5\},$$

where p_{vac} is the probability an employee is on vacation and p_{non} is the probability from the binomial distribution that an employee not on vacation is absent. Further, the callbacks need to be balanced across the jobs, so that

$$y_i - b_i \sum_{j=1}^{5} y_i \geq 0, \quad i \in \{1,2,3,4,5\},$$

where b_i is the percentage of callbacks that should be of type i. Finally, an additional constraint is needed to calculate the expected number of callbacks using the binomial distribution; hence

$$\sum_{i=1}^{5} y_i = E[Callbacks].$$

The expected number of callbacks is nonlinear in the decision variables . Indeed, if $M = \sum_{i=1}^{5} m_i$ is the total number of employees needed, and $N = \left\lfloor \frac{1}{2} + (1 - p_{vac}) \sum_{i=1}^{5} x_i \right\rfloor$ is the average number of employees not on vacation (rounded to the nearest integer) then

$$E[Callbacks] = \sum_{i=N-M}^{N} (M - N + i) \binom{N}{i} p_{non}^{i} (1 - p_{non})^{N-i}.$$

Because the problem is nonlinear, algorithms find locally optimal solutions, not globally optimal solutions.

The student team created a Microsoft Excel workbook using the built-in solver to find a candidate optimal solution (See Screenshot 2.3.1). Students created a simulation to model the callback process of the fire department (See Screenshot 2.3.2). The simulation, when fed the staffing levels from the candidate optimal solution, returns an expected cost similar to that predicted by the nonlinear solver and less than the fire department's current costs. These software tools were delivered to the fire department, who will use them to analyze proposed changes in vacation policy, callback policy, or required staffing levels. The candidate optimal solution has 374 filled positions, rather than the currently allotted 372 positions, of which 10 are vacant. Also, this solution shifts some positions (see Table 2.3.4). The simulation team predicts that this new staffing level would save approximately $250,000 per year.

Screenshot 2.3.1. The nonlinear solver.

Structure of the Project

Applied Mathematics Laboratory research projects are mutually beneficial for both our students and the project sponsor. Our projects provide experiences to the participating team members beyond those of traditional university courses. For the student members of the team, the experience is valuable for showing them the rigors and pleasures of research. It provides an opportunity for a student to use theoretical material from the classroom, new material from personal research, and the information and expertise of other team members in a cooperative educational setting. The processes of understanding the prob-

Table 2.3.4. Recommended staffing level from the nonlinear program.

	Optimal Positions	Required Personnel	Current Slots (3/2003)
Officers	61	61	62
Pump Operator	36	36	40
Vehicle Driver	39	39	40
Firefighter	225	144	220
Specialized	13	11	11
Total	**374**	**291**	**372**

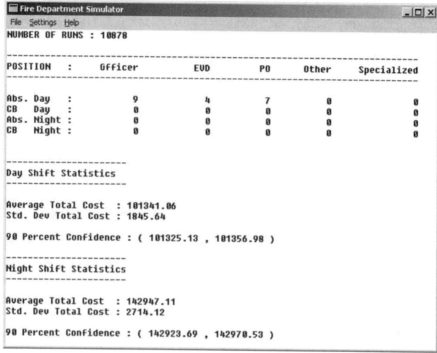

Screenshot 2.3.2. The simulation.

lem, assisting in the mathematical formulation, developing the necessary skills, obtaining information and sharpening expository skills are as important as the actual solution of the problem. In addition, the close working relationship that students have with faculty allows a personal dimension to their education which both parties enjoy. For our students this is a unique experience in learning: how to work with others, how to express complicated ideas clearly and confidently (in written and in oral form), and how to use the many skills and resources available to them.

There are clear benefits to the sponsor as well. The sponsor obtains insight into mathematical problems that demand time or expertise not available from permanent staff. Sponsors also benefit by having contact with some of Towson University's brightest students as potential future employees. Additionally, there is significant public relations value in supporting higher education.

Finding these partnerships, however, is no easy task. A great deal of personal networking is required. Another successful approach has been to have public lectures on applied mathematics and extend invitations to local government and industry. In these invitation letters, we describe the Applied Mathematics Laboratory and its mission.

This project began when we received a call from Robert Maloney, the Special Assistant to the Chief of the Baltimore City Fire Department, and a Towson University alumnus. He and Fire Chief William Goodwin were trying to control their overtime expenditures, and our letter came at an opportune moment. We arranged for a meeting between Robert Maloney and the Director of the Applied Mathematics Laboratory in early July. In this meeting we discussed the goals and objectives of the project. After discussions with the department chair, we decided to go ahead with the project. We then selected the project's faculty directors. After meeting with Fire Chief William Goodwin and his staff, we began to select the student participants.

We decided to seek six junior or senior mathematics students, some of whom had extensive computer programming experience. In early August 2002, we generated a list of all junior and senior mathematics majors who had at least a 3.5 GPA. We narrowed this list down using the students' academic records and our personal knowledge of the students. We then invited twelve students to an informational meeting in late August. Eight of these (four juniors and four seniors) indicated an interest in the project, and we selected all four seniors and two juniors. These students were invited to join the project and register for the Applied Mathematics Lab I in the fall of 2002. All six students chose to continue with the Applied Mathematics Lab II in the spring of 2003.

The Applied Mathematics Lab I and II met twice a week for one and one-quarter hours. We felt that it was important to have several formal meetings each week. This schedule was designed so that the students could be given small tasks to work on for one half of the week in the hope that this would keep them constantly working on the project instead of having large amounts of time with nothing to do. The first few weeks of the course were devoted to an introduction to mathematical optimization and discrete event simulation. This introduction gave the students the basic background material. Once that basic instruction was completed, the students were broken up into two teams of three. One team explored the optimization problem, while the second team considered the simulation.

One advantage of this structure was that it was possible to take one member from each team to form a third team of two to solve a new short-term project. The remaining team members continued to work on the core project. For example, this procedure was used to create a team that wrote a successful small grant proposal to the Towson University College of Science and Mathematics Committee for Undergraduate Research.

At the end of the first semester, the students produced a written interim report and gave a presentation to fire department officials. At the conclusion of the project, the students presented their results to the leadership of Baltimore City. They also produced a final written report and created two software tools for the fire department.

Challenges

This project was not without its difficulties. In each semester the students received three credits for the Applied Mathematics Laboratory course, but they spent more time on this course than they would have for a normal three credit class. To alleviate the problems with this time commitment, we encouraged the students to take lighter than normal course loads. The students also had to deal with the problem inherent in all research; sometimes the chosen methods simply fail. The students tried several approaches to the problem before finally finding the ones that produced reasonable results. Two additional challenges can be identified. First, there are issues with the group dynamics of the student teams. Second, as in any real world modeling problem, there are difficulties in acquiring meaningful data in usable form.

The student team was a group of academically high achieving students. However, these students came from a wide range of backgrounds and did not form a cohesive unit. Since we knew very little about any of the students when the project began, we allowed them to choose which team they would like to work on. In addition, we tried to let the teams delegate the required tasks as they saw fit. This led to a major problem, since we had several extremely motivated and diligent students who took it upon themselves to do most of those tasks and left the rest of the students with nothing to do. The other students, while able to perform any task given to them, allowed those motivated students to do most of the work.

As the project advanced and this problem was noted, the directors directly assigned projects to the students with less work. This method seemed to work well. All of the students were highly capable, and this allowed all of their skills to be used. Unfortunately this procedure was not implemented initially, as it would have greatly improved the group dynamics.

Any real modeling problem has difficulties with the data. One notable problem was that the data on the callbacks and absences for the department is not complete. The callback data was listed by shift and rank (captain, lieutenant, etc.), but the absence data was broken down only by shift and cause (vacation, illness etc.). This made it difficult to determine the relationship between the absences and the callbacks. In addition, this relationship is highly dependent on the number of people scheduled for a given shift. Further, the staffing level was not constant over time. In fact, the best we could do was to estimate the staffing level at a given time based on several approximations.

Similar difficulties have occurred in previous Applied Mathematics Laboratory projects. However, these difficulties did not interfere with the successful accomplishment of the project goals.

Conclusion

This project was successful in a number of ways. The students had the opportunity to do undergraduate research on a real project of keen interest to the fire department. As the project's capstone, they presented their work to the top officials in the City and the fire department at an official briefing. The unanticipated news coverage of this event highlighted the fact that even though mathematical research for the public sector may not be as profitable as in the private sector, there are other intangible benefits. The fire department received an analysis of their current staffing model and software to analyze changes in their policies. These are tools that they were unable to produce internally. The project was of great benefit to everyone.

Acknowledgements

First we would like to thank our students Kristina Diaz, Michael Machovec, Péguy Pierre-Louis, Marco Radzinschi, Melissa Rosendorf, and Ian Stalfort. This project would not have succeeded without their outstanding efforts.

The authors would like to thank Martha Siegel, our department chair and former Director of the Applied Mathematics Laboratory, for all of her help. We would also like to thank Dean Gerald Intemann who supported us with an undergraduate research grant. We thank Fire Chief William Goodwin for providing the opportunity for this project.

Special thanks go to Robert Maloney, who acted as the fire department's liaison and provided us with the data needed to solve the problem.

References

1. http://www.ci.baltimore.md.us/news/citistat/
2. Fox 45 (WBFF), 10:00 news on Friday, May 8, 2003.
3. http://www.thewbalchannel.com/education/2194251/detail.html

About the Authors

Andrew Engel joined Towson University in 2002. He received his PhD from The University of Arizona in 2002. His research interests include game theory, dynamical systems and operations research.

Coy L. May is a professor of mathematics at Towson University. He received his PhD from the University of Texas in 1975. His primary research interest is group actions on surfaces. He has directed several Applied Mathematics Laboratory projects over the years; three of these projects have involved scheduling problems.

Mike O'Leary came to Towson University in 1998, and became director of the Applied Mathematics Laboratory in 2000. His research interests include fluid dynamics and partial differential equations.

2.4

Creating Service-Learning Experience in an Experiential Learning Environment

John R. Ramsay
The College of Wooster

Abstract. The College of Wooster Applied Mathematics Research Experience is a summer program that employs students to work as consultants in the local community. Students generally work in teams of three with a mathematics or computer science faculty member acting as advisor. Clients of the program come from business, industry, government agencies, and service organizations. The program also includes a significant service-learning experience in order to increase the educational benefit to the participants.

Introduction

The variety of different service-learning activities is almost equaled by the variety of ways in which service-learning can be defined. Most typically, these activities occur as specific service-learning components within an academic course, but they can occur in many other ways as well. One interesting way to provide service-learning experience for students is to introduce it as part of an existing work or internship program. Students can work within the program community, attend seminars and workshops, and spend time analyzing the strengths and weaknesses of the work environment in which they are employed.

Each of the above elements corresponds closely to the components described in most definitions of service-learning: involving students in service, building a sense of community, relating the work to the academic curriculum, and providing structured reflection on the experience. (See, for example, Campus Compact [2].) Identifying such components is important in providing ways to distinguish service-learning from other similar experiences such as volunteer work, internships, or experiential learning. Such a definition can be applied to many programs and activities that cover a very broad spectrum of experiences.

A look at the service-learning philosophy among the so-called "work colleges" of America, schools which require all students to work in some capacity as part of the educational program, reveals service-learning not so much as a specific activity to be implemented in a particular place in the curriculum but as an educational philosophy that permeates the entire academic experience. Of course not all institutions are work colleges (nor should they be), so service-learning will generally occur on a much smaller scale and usually only in certain parts of academic programs. However, this broader view of service-learning as an educational philosophy allows service-learning opportunities to be identified in a greater variety of programs and activities. In this context, service-learning spans the middle of a spectrum of service programs with volunteerism (recipient service) on the one side and internships (provider learning) on the other [3]. With this model in mind, one can more easily identify service-learning occurring within or as a component of a program that includes other experiences and has other goals as well. At The College of Wooster, we have such a program.

The College of Wooster AMRE Program

Overview

The Department of Mathematics and Computer Science at Wooster has an eight-week summer program that provides mathematics and computer science students the opportunity to work for business, industry, and social/civic organizations as part of an educational consulting agency. The program, The Applied Mathematics Research Experience (AMRE), was created primarily as a means to give mathematics and computer science students at Wooster a taste of real world applications of their academic work. We determined at the outset that we wanted the program to be both an educational experience and an attractive summer option for our students. Though our students value the educational and experiential opportunity the program provides, summers are a time when most of them need to focus on the financial needs their college study creates. For this reason, the program is designed to be a summer job for the students rather than to carry academic credit. Very few of our students need additional academic credits.

A typical AMRE summer includes three or four teams each working on a project for a client. Most often, a team works for one client, though occasionally a team will work for two clients on related projects. (For example, one of the teams for summer 2003 did a market analysis and demographic data collection for two separate organizations: the local arts center and the corporation of downtown businesses.) One AMRE team usually consists of three students and one faculty advisor.

AMRE includes many educational activities that involve all students in the program. These include workshops on consulting, presentations and communication, occasional lectures or colloquia on topics related to the students' work, and regular presentations and discussion of the work being done. Further, department faculty are involved on a daily basis advising students on specific directions and tasks as well as furthering the educational experience of the students. A fall AMRE Projects Day provides an opportunity for the student teams to present their work to their peers, college faculty and administrators, and potential future clients.

History

The Mathematics and Computer Science Department created the AMRE program during the summer of 1994. That first summer included projects for the City of Wooster and for a local dairy company. Each subsequent summer has involved between five and twelve students working on a total of three or four projects.

AMRE was initiated with support from a seed grant from The College of Wooster's Hewlett-Mellon Presidential Discretionary Fund for Institutional Advancement, and continued support from this fund was provided through 1997. Additional funds came from the College's Office of Undergraduate Research and the Office of the Vice President for Academic Affairs. A fee from each of the program's clients has been an increasingly important part of the funding. Fees in the first two years were small (one or two thousand dollars) but important to guarantee that the client had some investment in the success of the project.

Program Administration

In order to seek potential clients, solicitation letters are sent in January by the AMRE program director to area firms and organizations. The director follows up with phone calls, more phone calls, and more phone calls. Department faculty review potential projects and accept several into the program. The number of projects chosen depends on the quality of the project proposals, the number of faculty available to advise teams, and the pool of student applicants. A faculty advisor is chosen for each accepted project and an informal *Letter of Understanding* provides the contractual agreement between The College and each client organization. The client organization identifies a liaison as the primary contact for each AMRE team. In the spring the faculty advisor for a particular team works with the client liaison to determine the scope of the project and formulate a more detailed project statement. The director and faculty advisors select the AMRE students from among applicants to the program. Approximately one third of the applicants are offered positions. Most participating students are just completing their sophomore or junior year, and though they are usually mathematics or computer science majors, some have come from majors including physics, economics, chemistry, international relations, and religious studies.

Program Funding

Currently the cost of each project is about $18,000. Approximately $11,000 goes to the stipend, room, and board for the three students. The client fee covers about one-half the cost and the College provides the rest, primarily through an endowed fund established in 2000 as part of an alumni class gift specifically for this program. Additional support is also provided through a summer research fund earmarked specifically for students with sophomore standing.

Two AMRE projects

To illustrate the sort of projects undertaken by the AMRE program, two examples are discussed below. These two projects have been chosen because they provided service to a community extending beyond the particular client organization.

Figure 2.4.1. AMRE students are shown specific pests being studied at the Ohio State University Agricultural Research and Development Center.

Interactive Web-Site Enabling Ohio Growers To Obtain Plant Phenology Information To Be Used In Pest Management Decisions

The AMRE team created an interactive web site to enable a grower anywhere in Ohio to input his/her 5-digit zip code and obtain information as to which insects/pests are likely to be present at the time (Figure 2.4.1). The growers can use this information to determine which particular pesticide sprays should be applied. Most growers spray their nurseries every five days with multiple pesticides in order to insure that the one relevant to current insect or weed pest presence is used. In an effort to address the environmental concerns created by this over-spraying, the US government has been funding research that might reduce the number of sprayings. Over a period of three years, Ohio State University's Ohio Agricultural Research and Development Center (OARDC) monitored the phenological (growth and development) sequence for 85 plants and 46 insect and mite species and compiled a database linking growing degree-day and this phenological data. The easily monitored sequence of flowering plants was used as a biological calendar to track degree-day accumulation and predict the phenology of insect and pest weeds. Growing Degree Days (GDD) are a measure of the growth and development of plants and insects during the growing season. The underlying idea is that development does not occur unless the temperature exceeds a certain minimum threshold value or base temperature (Figure 2.4.2). The base temperature used for each organism is determined through research and experimentation.

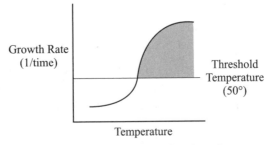

Figure 2.4.2. Growth occurs as a function of temperature above a "threshold."

Using daily weather information for a particular region, an accumulation of Growing Degree Days is computed and used to make reasonable estimates of plant and insect growth patterns during the growing season.

Once the underlying database was created, mathematical and computer science skills were needed to make it available to meet the needs of Ohio growers. To this end, the AMRE project was divided into the following two components:

1. Devising a mathematical formula that uses temperature data from ten weather stations positioned around the state to approximate the Growing Degree Days value for any zip code location.

2. Creating an interactive web page allowing the grower to access the appropriate phenological data.

To accomplish the first task, the AMRE students wrote a computer program that applies a weighted average of the GDDs of ten state weather stations to estimate the GDD of any zip code location in Ohio. The model that the program implements uses the latitude and longitude of the zip code location entered and that of the ten OARDC weather stations located around Ohio (Figure 2.4.3) to compute a "distance" between the site and each of the ten weather stations. The distance formula used is weighted to include factors such as the difference between North-South and East-West variation in temperature. The program then uses these individual distances in conjunction with the GDDs of the ten weather stations to estimate the GDD of the site.

Figure 2.4.3. Location of weather stations used for interpolating Growing Degree Days for other points in the state.

The students designed a web page that implements this GDD program and interfaces with the database of lifecycle pattern information of indicator plants and insects. The web site [4] is able to make a fairly accurate estimation of all the phenological events that are likely to occur on or around a particular day in the general area of the entered zip code. This information helps growers make better and more environmentally sound pest management decisions.

Web Database Project to Facilitate Collection and Retrieval of Language Sample Data

This project was commissioned by the Augmentative and Alternative Communication Institute (AAC), a non-profit organization started by the Prentke Romich Company. Prentke Romich manufactures augmentative communications devices, enabling persons with disabilities to communicate more effectively (Figure 2.4.4). One of the activities of the AAC Institute is to administer a web site as a service to those in the Augmentative and Alternative Communication field. The primary purpose of the AMRE project was to enable the web site users (primarily voice or speech pathologists) to view language samples and their corresponding analysis.

Communication devices equipped with certain features have the ability to record types of interactions between the user and the device. These interactions fall into three usage categories: full word spelling, word prediction and semantic compaction. (Semantic compaction is keyboard use that takes advantage of an elaborate language programmed into the device that is implemented using specially designed keyboard overlays.) The AMRE

Figure 2.4.4. AMRE students are given demonstrations of the communication devices.

students wrote a web interface that allows users to access interaction information stored in the database. By analyzing this usage data, individuals in the AAC field can better understand the communication patterns of device users. The data can be distilled to provide various useful summary statistics, such as the number of utterances, the average length of an utterance, or the communication rate of the user within each usage category. Since the interface is web based, information can be retrieved from virtually any computer with a web browser.

In order to facilitate useful queries for language samples, the system designed by the students stores personal characteristics of every subject who has language samples in the database. This personal information is divided into static information, such as name and birth date, and potentially changing information, such as living status, marital status, education, and employment. These variable attributes together form the profile of the subject; since the attributes may vary with time, there may be multiple profiles of the same subject. Further, there may be multiple language samples for a particular subject under a particular profile. By having access to the data provided by the web site, speech and voice pathologists are better able to assist clients in making decisions about device appropriateness and device use.

Work and Learning

At the work colleges, service, work, and learning are each part of the overall educational experience. These institutions integrate academics, work, and service as part of the college mission; each influencing the others in all aspects of the student experience. One need simply look at the language of their mission statements to see this integration. At Warren Wilson College, students practice "an integrated triad of intellectual study, useful and productive work, and service to others beyond the campus community, thus providing opportunities to develop the capacity for lifelong work, an effective career, and service" [6]. Blackburn College is about the business of "Providing 'hands-on' work, service and leadership learning opportunities in addition to the classroom" and "providing an added dimension of community involvement and student character development emphasizing a strong work ethic, responsibility and accountability" [1]. Work is viewed as a part of student development and as part of living and serving in a community. It is interpreted in terms of responsibility and service to a larger, common good.

The AMRE program was designed with a similar view. The overlap between the work experience and the service-learning experience is sufficient enough that there are no clear lines to draw within the program. The various experiences are interrelated in such a way that identifying some portions as "work," others as "service," and yet others as "academic" doesn't reflect the nature of the overall program.

In his theme paper for the 1987 Work-School Conference, William Ramsay, former Vice President for Labor and Student Life at Berea College, identifies the following five "benefits" these institutions claim to make for their programs of service-learning: financial aid benefits, educational benefits, societal benefits, career development benefits, and community service. At first blush, one might balk at the inclusion of financial aid (pay for work performed) and career development. But when service-learning is part of a larger experience, these benefits can be consistent with the goals of a service-learning experience. Students are able to be part of an experience in which they "are not simply consumers of educational services but are contributors" [5]. They become part of the broader community and accomplish tasks and produce goods that are needed in this community and the society they live in.

However, it should be noted that doing productive work by itself does not meet the requirements of service-learning. Otherwise, experiential learning opportunities such as internships and co-ops would also be included. Service-learning experience occurs when this work is done in the context of a program committed to relating the work to the educational process and to reflecting on the value of work and contribution to community. This is what we have tried to accomplish in the AMRE program.

Service-Learning and AMRE

Though primarily a work experience, AMRE includes many service-learning experiences for the students. As should be clear from the above description of the program, there are certainly financial benefits to the student as well as obvious benefits in career development. It is in other areas that I would like to illustrate how service-learning experience occurs within the AMRE program.

Educational Benefits

Applying specific mathematical skills as well as more general problem solving skills developed by studying mathematics reinforces and legitimizes knowledge learned in the classroom for participating students. The opportunity of experiencing first-hand the value of the skills they are developing is important in the building of self-confidence and the exploration of future educational goals. Blending the experience with educational activities within the program further strengthens the connection between their academic experience and the work experience. An important way to establish these connections for the students in the program has been to directly tie the work they are doing with specific activities that are more familiar to their usual academic experience. Early in the program there are usually specific areas of knowledge that a particular team of students need to acquire in order to effectively address the project at hand. In these instances, one of the faculty advisors will provide a lecture or a series of lectures containing the needed information. These lectures are attended by all of the students in the program so as to provide for an educational experience that is as broad as possible. For example, lectures have been given on artificial neural networks, production scheduling, statistics, and economic order quantity modeling.

On an annual basis we have a professional consultant come to campus to give the participants a workshop on consulting and teamwork, and a member of The College's Career Services Office gives a session on oral presentations. Oral presentation skills are further practiced through weekly presentations given in-house. Midterm and final presentations to the client are required (Figure 2.4.5).

Societal Benefits

A particular strength of the program is giving students an opportunity to be contributors to society. Most students are at a stage in their lives where they are primarily users of the ben-

Figure 2.4.5. Midterm and final presentations to AMRE clients are required components of the program.

efits of society or, at least, users of society's educational system. The AMRE program gives them an opportunity to be part of the societal work force. Seeing themselves as an important part of a larger sphere of individuals gives them a sense of belonging and a sense of responsibility toward something outside of themselves. The experience also develops initiative and leadership skills that will influence the participants as they become active citizens.

Community Service

The AMRE program provides a community experience at two different levels. The teamwork and group activities create a community within the program that enables students to be equal partners in a work experience that has community and team goals rather than individual goals. They learn early in the program that they are better as a team than they are as a collection of individuals. This develops important understanding of the value of community and the need to build community. Another level of community experience is in involving students in the local community. One of the difficulties we have as a relatively small, private, national liberal arts college is the divide that occurs between the campus community and the surrounding community. Having students work as equals with members of the community, whether it be service organizations or business and industry, creates bridges between the two groups and gives the students a sense of belonging in the larger community. This is an important first step in helping students understand the need for involvement in issues that occur in this larger community.

In several cases, the projects themselves reflect work done very specifically to serve the local community. The two examples described above were such projects. In both cases, the funding came from grant money supporting a particular non-profit organization's attempt to address the needs of a particular community. A number of other projects of this nature have occurred as well. One could envision creating a program whose projects were exclusively with service organizations, but it would require substantial effort to find enough clients and sufficient funding.

Reflection

One of the most important aspects of a service-learning experience is reflection on the work performed. The AMRE program provides numerous opportunities for the students and faculty to reflect on their experiences. This occurs in a variety of ways, some informal, some formal. The weekly in-house presentations provide opportunity for the full group to reflect and discuss the work and experiences of each of the teams. Further opportunity for this is provided with weekly lunches that bring all students and faculty together for informal conversation. It is in these sessions that most of the opportunities for reflecting on aspects of the business environment occur. With students working in a variety of different settings, it doesn't take any prompting to begin discussions that compare the different environments. Students see contrasts in management and leadership style and differences in worker morale. Students are able to identify the actions and attitudes that create a positive business environment through reflection on their exposure to businesses that treat their employees with dignity and respect as well as businesses that don't.

One of the first businesses with which we worked was a dairy. After a couple of visits, the students commented that the environment was different from what they had expected. The employees at all levels seemed happy and enthusiastic about their work and it was not always easy to tell who was working at a blue collar job and who was management. After they made these comments, we took time to look at the mission statement of the company and it was clear that the goal of treating everyone with "love, dignity and respect" creating a "challenging and rewarding work environment" were not just words. It was satisfying to hear the students identifying specific things they had seen done in order to create the environment they had experienced.

A very similar experience occurred in another project at a company that listed as part of its mission to create a secure and stable employment opportunity for the overall community. Students learned of conservative investment and expansion decisions that were made in order to guarantee stability as a top priority rather than produce the greatest profits. As the team was taken on a tour of the grounds, we observed the company founder and recently retired CEO on his tractor where he continued to work the family farm. One of my students, making an immediate connection to the operations research course he had taken with me the previous spring, commented, "This is one of those cases you talked about where 'maximization of profits' is not the primary goal of the company, isn't it?" I couldn't have said it better myself.

Students have learned from negative environments as well. Reflective comments about employees "looking over their shoulders" and managers unable to communicate with others have also provided important learning experience.

Another informal opportunity to reflect on work done occurs in the fall following the summer program. Each student team presents their work to the college and local communities at an annual *AMRE Projects Day*. The client liaisons from the previous summer attend the session and have the opportunity to discuss the outcomes of the students' work. This further strengthens the sense of responsibility and satisfaction that comes with having provided a service to a part of the larger community.

There are two more formal reflection exercises that are required of students in the program. In the last week of the program, each student is required to provide a written evaluation of the experience. Further, students at sophomore status are required to write a report on the experience that is submitted to the organization that provides funding for those positions.

Conclusion

Universities and colleges divide into academic disciplines in order to focus on the fundamental ways of learning in particular fields of study. Interdisciplinary endeavors complement this division of experience as students progress through the curriculum by offering one way to better relate the academic experience to the more complex world in which we live. Service-learning projects have the potential to offer this kind of broader experience that can help students interrelate many different aspects of their education. Often, the experience occurs within a specific course with service-learning as the center of the desired outcome. In other cases, we can include service-learning experiences as part of larger, more comprehensive endeavors. It is hoped that seeing examples of this latter model will encourage academic institutions to find ways of incorporating service-learning into their programs.

References

Blackburn College *Mission Statement*. Retrieved June 18, 2003, from www.blackburn.edu/wp_mission.statement.asp

Campus Compact (2000). "Definitions of Service-learning." Introduction to Service-Learning Toolkit, pp 15–17.

Furco, A. (2000). "Service-Learning: A Balanced Approach to Experiential Education." Introduction to Service-Learning Toolkit, pp 9–13.

OARDC, The Ohio State University, *Growing Degree Days and Phenology for Ohio*. Retrieved June 18, 2003, from www.oardc.ohio-state.edu/gdd

Ramsay, W. (1987). "Student Work Colleges: A Service-Learning Philosophy and Style." Work School Conference on Service-Learning. Berea College.

Warren Wilson College *Mission Statement*. Retrieved June 18, 2003, from www.warren-wilson.edu/catalog/mission.shtml

About the Author

John Ramsay is Professor of Mathematics at The College of Wooster. He received his BA in mathematics from Berea College and his MS and PhD from The University of Wisconsin, Madison. Current areas of professional interest are American and Irish calculus instruction and undergraduate research. He is founder and director of The College's Applied Mathematics Research Experience.

2.5

The Mathematics Umbrella: Modeling and Education

Arcadii Z. Grinshpan

University of South Florida

Abstract. The "Mathematics Umbrella" program is aimed at bringing active experiential learning into the curricula of some collegiate mathematics courses through an optional project that can substitute for some course requirements; it also forges a mutually beneficial partnership between mathematics professors and the non-mathematical community along educational lines. The ideas presented here are based on the long-term usage of interdisciplinary and business projects in engineering, life sciences, and business calculus courses at the University of South Florida. One such project, a transgenic mouse population model, is discussed in detail.

Introduction

Mathematical language is abstract. To a great number of university students, traditional mathematics instruction seems to be much too theoretical: formal lecturing and abstract exercises. Perhaps they are not entirely wrong, as collegiate mathematics education sometimes tends to be quite distanced from the real world and often ignores students' professional interests. Such a framework does not allow students to extract the maximum benefits from the mathematics. Non-mathematics majors, the vast majority of the student body, simply have to put up with this situation. Most of them study mathematics just because they have to do it. Regrettably, quite a number of these soon-to-be professionals are unaware of the importance of mathematics in their prospective fields.

For years there have been numerous attempts to improve collegiate mathematics education for non-mathematicians. Bringing active experiential learning into the curricula of some mathematics courses is supported by many positive instances and seems to make the learning process work. Focusing on real world problems is of priority because those problems provide the most benefits to mathematics education. Also it boosts students' interest in serving the community (cf. [2, 3]). Taken in this context it is natural for us to encourage undergraduates involved in any professional activities or those who are in the midst of studying the major disciplines to apply mathematics to their fields of interest. In this case, the necessary help from both mathematicians and non-mathematical professionals is widely available in many university communities, where service-learning activities are especially welcome.

As a means of relating mathematics education to the needs of the community, including our own university community, as well as to create specific service-learning opportunities for students, we sponsor a program called the Mathematics Umbrella. This is a structure that allows for an active collaboration of mathematicians with other specialists and is based on a broad involvement of students. It is oriented toward current business/science problems and makes it possible for mathematics-business/science projects produced by students to become a component of their collegiate mathematics education. Such a mathematics umbrella program could be adaptable to a number of university environments. In our own implementation of this concept, not all the typical components of a service-learning program are present (e.g., a formal reflection process), but the framework is somewhat novel and we see clear potential for a full service-learning experience.

In this paper I share my experiences in launching the Mathematics Umbrella Group (MUG) at the University of South Florida (USF), which is located in the Greater Tampa Bay area. An essential characteristic of MUG is its inclusiveness. It provides a mechanism for a flexible collaboration among USF students, USF professors from different fields, and various community professionals. The problems addressed come from different organizations, but all of them require mathematics. So far MUG has been engaged in this activity for five years and it seems to have had a positive impact, even though there are continuing attitudinal and structural challenges both inside and outside the university. The MUG website [5], developed jointly with David Milligan (College of Education, USF), shows its structure and contains both educational and applied mathematics information.

Special appreciation for support and involvement goes to Marcus McWaters, chair of the department of mathematics at USF, as well as to dozens of USF professors and other experts from Florida who have proposed fascinating problems based on real data and helped me to supervise the student projects. They represent different spheres of activities such as athletics, business, communications, entertainment, finance, government, industry, medicine, public health, sciences, and transportation. Among them are engineers, lawyers, managers, physicians, police officers, as well as professors of anthropology, biology, chemistry, computer science, environmental science, engineering, geology, medicine, psychology, and physics. My applied mathematics horizons have been greatly expanded in the course of this activity. I would also like to thank all the student contributors for their work.

Calculus Courses and Real Life Problems

Mathematics professors traditionally emphasize comprehensive problem-solving competence. However, there is generally quite limited time for non-mathematics majors to devote themselves to collegiate mathematics, and applying mathematics in a friendly environment is more appealing to them than developing problem-solving skills by solving tricky, abstract problems. Therefore, as the trend in many modern textbooks for elementary courses suggests, we might well do better concentrating on mathematics applications in students' future professions and in this way we persuade them that mathematics education can be quite beneficial (cf. [3]).

To narrow the focus of our discussion, consider the case of calculus. Calculus students are usually grouped into several course categories such as engineering calculus, life sciences calculus, and business calculus, with a variety of distinct majors in each group. In addition, some elements of algebra, differential equations, discrete mathematics, and probability/statistics might be incorporated into some of the calculus curricula, thus adding further breadth. Nevertheless, because of the diversity of student interests and the relative abstractness of the calculus curriculum, by itself this approach is still limited in helping the students connect the subject of calculus to their major academic and professional interests. Since the prospect of distinct calculus courses to suit many majors or areas of interest seems quite impractical, we have pursued with MUG an alternative framework to give students an opportunity to connect their mathematics study with their own interest areas. This framework consists of allowing students in certain courses the option of pursuing a suitable project through MUG in place of certain course requirements, including, under certain conditions, the final exam.

One necessary condition for this process to function successfully is the presence of mathematics professors who are able to work with diverse professionals in the community as well as with professors in other departments. We can think of each project as a triangular structure, the sides of each symbolic triangle uniting a mathematician, another professional, and one or two students. It is not unusual if one mathematician is involved in several projects. Some interesting problems can further be reframed to be of use for educational purposes on different levels. The mathematics umbrella approach at USF [5] is just one example of how to pursue this concept. Similar mechanisms may be established in different schools, possibly in different ways. If this is done, both the community and mathematics curricula can be positively affected.

Let us look at the project option, which is an innovative feature of some credit-bearing mathematics courses designed to allow students to participate in experiential learning outside the classroom. Students are in no way obligated to take advantage of this opportunity, but this can be an excellent way for them to make tangible connections between mathematics (calculus, in particular) and the physical world. The role of the non-mathematical professionals is of course quite important, for they are the ones who propose and consult on student projects. Since 1999, the Mathematics Umbrella Group at USF has been supporting such a project option for non-mathematics major students, generally in calculus.

Which students are interested in the project option? By the look of it, it seems that the project option is of particular interest to those students who are already involved in business or research: company/organization employees, research assistants, honors students, and interns. Natural leaders are inspired by the idea of a project and jump at the opportunity to learn in a practical setting. The same is true for the students who are good at mathematics applications and prefer them over solving formal mathematics problems.

What problems are welcome and how do we match them with course curricula? A project should stem from current or planned organization (company, department, government) activities that require mathematics applications. Once a student taking a given mathematics course (e.g., calculus) is interested in a particular project, a mathematics professor (generally the industrial mathematics coordinator or course instructor) needs to bring the mathematical side of the project into alignment with the course curriculum, provide advice on mathematical tools, and assist in resolving unconventional cases or problems.

How do we grade a project and give a course grade when the project option is elected? The student's level of accomplishment, the project's mathematical content, and the project's usefulness to the receiving organization all enter into a project grade. Of course, this is a delicate and judgmental matter. It becomes easier when a system is already in place and it is possible to judge the project in comparison with the similar ones completed earlier. But then there is also the issue of how to weigh this grade in the overall determination of a course grade. For example, allowing a student's project grade to replace a lower test grade is one policy that is frequently followed. In this case the weight of the project grade will be greater if fewer tests are given in a course. I have used several schemes in attempts to design a proper grading system. All of them have two features in common. First, the influence of the project grade upon the final grade should be significant. Second, the project option and its assessment should not push the basic course content aside. To achieve these objectives, I sometimes divide students into three categories S, P, and U. Here, S/U (satisfactory/unsatisfactory) means that according to the test scores a student has/has not demonstrated a working understanding of the course concepts. P is an intermediate category indicating partial understanding. I allow "S"-students to have their project grade (which will tend to be high for them) as their course grade. For "P"-students the project grade may count just for a portion of the final grade. Thus projects become a gate to a better grade. I find it reasonable that project option students have their final exams waived if they wish, since I prefer a grading system that highlights both the practical ability to use mathematics and the conceptual understanding of course material.

What benefits does the project option have for promoting our educational objectives? According to my experience, the project option seems to be a good way for students to develop a better sense of mathematics as it is applied to an area they are interested in. Mathematics-business/science projects involve a synthesizing experience that requires a better and fuller kind of understanding of mathematics in context. Mathematics may become less intimidating to students after this kind of experience. Another advantage is that students develop an awareness of the diversity of mathematics applications, which they encounter in the process of picking a project. In addition, projects give educators a chance to communicate with individual students more closely. Providing students with individual discussions and learning sessions stimulates them to apply mathematical tools more thoroughly and enhances their position for later career opportunities. In addition, educators learn of new applied ideas, which can be incorporated into other parts of the mathematics curriculum.

Now let us have a look at how the USF program operates. The project option has become popular as a part of our calculus curricula, generally being an alternative to the traditional final examination. The mandatory tests aimed at checking conceptual understanding are the same for all students. Thus, if a student does well enough with those tests, then she or he faces two options: a final examination involving standard formal calculus problems, or a mathematics-business/science project that applies mathematics to a real life problem. Over two hundred fifty students have been involved in project activities and more than two hundred twenty projects have been completed over the past few years. My engineering and life sciences calculus students have done the most projects. Other projects have been executed by David Milligan's business calculus students and by Fred Zerla's and Sam Isaak's life sciences calculus students. Community professionals provided about 70% of the problems for projects, while USF professors proposed the remaining 30%. The mathematical aspect of each problem has been discussed with me. Thus, each project is the fruit of efforts carried out under the joint supervision of a non-mathematician (i.e., a USF professor or company/organization expert who proposed the problem) and myself. Student projects deal with applications of mathematics to the problems of hospitals and clinics, industrial and financial companies, supermarkets and restaurants, university divisions, government agencies, and other organizations. (As the reader probably already realizes, although students often do not, mathematics happens to be applied everywhere.) I provide every opportunity for student contributors to take their time to do a project. There are no size restrictions either. In some cases two students produce a joint project. Although the basic requirements are informal, students must comply with the instructions in "Organization of project write-up," an annually renewed form.

The abstracts of completed projects can be found at the MUG website [5]. They are assigned to one of five categories: Human-Human, Life and Earth Sciences, Engineering, Economics-Finance-Banking, and Commerce. A good number of new projects are underway all the time.

From Simple Questions to Challenging Models

As mentioned above, the scope of problems allowed for student projects is quite wide. While some problems look simple, perhaps even trivial, for mathematicians, others turn out to be very difficult. Thus, not only do mathematics-oriented problems intrigue and challenge students when they work in the area of their specialization, but also problems posed by non-mathematicians are occasionally of genuine mathematical interest. The mathematical content of projects ranges from exercising standard formulas to using techniques from advanced calculus. Actually, the models developed in the student projects span essentially all basic concepts of college mathematics. The mathematical difficulties are worked out in our discussions with the student contributors, and some models require quite advanced advice on the subject.

Just taking biomedical examples out of the much wider variety of completed projects provides good illustrations of their interdisciplinary nature and curricular impact. For example, several students have applied mathematics to current genetic research, which requires what are called *transgenic mice*. These mice are used in models for human diseases, such as Alzheimer's, to test potential treatments (cf. [6, 7]). For instance, Jessica Maloney, a junior in biology, used calculus to study a genetic mouse mutation. Her project, "Mathematical analysis of the pathological markers of Alzheimer's disease," involved the Alzheimer's Disease Research Laboratory at USF and was supervised by its head, Dave Morgan.

The project I would like to use as an illustration is "Transgenic mice population growth," an interesting case proposed by Fiona Crawford. She served as the associate director of the Roskamp Institute, advancing the field of Alzheimer's research with her

Figure 2.5.1. "If you think your car is expensive, try taking care of lots of little mice!"

colleagues. (This institute is now located in Sarasota, Florida.) The transgenic mice involved in this project are very expensive animals, due to the fact that they must be genetically engineered (Figure 2.5.1). The focus of this investigation was to decide whether it would be feasible, given budget limitations, to purchase two parent mice for $11,000.00, one male and one female, and raise a population of mice of a specified size. To solve the problem we should have the time equation or algorithm to determine the amount of time needed for the resulting population of mice to number a given quantity. Then we figure out if we can afford the resulting population of offspring for this amount of time.

Brandon Faza, a junior in honors microbiology who was taking my engineering calculus course, was a student contributor to the project. He pointed out the facts that are needed for an idealized mathematical description of the problem (cf. [4]). For example, this species of mouse (Mus) lives for approximately two years, has a gestation period of approximately 19–21 days, and requires a period of eight weeks to reach sexual maturity. The litter size is generally 5–7 babies for this particular laboratory species, but it can reach up to 20 for some species of mice. Using these data Brandon assumed the average gestation period of 20 days, average litter size of 6 babies (3 males and 3 females), and a requirement of 60 days for sexual maturity. For simplicity he ignored the fact that some mice can die of natural and other causes within the given time (including the cases when reproducing females die in the end of the sixth reproduction period).

Let m_n be the number of mice and f_n be the number of females reproducing at time. The time unit is 20 days, thus $t_n = 20n$ days. Also we have the following initial conditions:

(1) $m_0 = 2$ (two parent mice) and $f_0 = \mathrm{f}_1 = f_2 = 1$.

Having found a non-trivial approximate solution to the problem, Brandon excitedly came to me with the words: "If you think your car is expensive, try taking care of lots of little mice!" His approach was based on an exponential approximation of the form $m_n \approx re^{\rho t_n}$. He used some numerical analysis to determine the suitable values of r and ρ, provided that several initial members of the sequence $\{m_n\}$ were known. This is an interesting practical strategy, though it cannot fully serve its purpose due to poor convergence. In particular for $r = 2$, the case considered by Brandon, there is some deficiency. A computational algorithm based on recurrence relations or a closed formula describing the size of the mouse population as a function of time under the above conditions would be expected to fit better here. One might consider the Fibonacci sequence and his rabbit population model as a hint for a student. However the mouse population problem still requires a mathematician's eye on recognizing and handling recurrence relations.

Built upon Fibonacci numbers and innovative efforts to cure Alzheimer's disease, this problem and its variations greatly appeal to students. Their enthusiasm and importance of the subject matter encouraged me to devote more attention to sequences, polynomial equations, complex numbers, trigonometric formulas, and approximations within the calculus course in order to better prepare students for these types of problems. In this connection the two following approaches based on recurrence relations for sequences $\{m_n\}$ and $\{f_n\}$ are worth mentioning.

First of all, it is easy to see that these sequences are related by the equation

$$(2) \qquad\qquad m_n = m_{n-1} + 6f_{n-1} \quad \text{for } n = 1, 2,\dots$$

Then we note that the sequence itself satisfies the recurrence relation

$$(3) \qquad\qquad f_n = f_{n-1} + 3f_{n-3}, \quad n \geq 3.$$

One can use (1), (2), and (3) to determine a number of successive values of both sequences and thus step-by-step solve the considered problem computationally. Here are some values of f_n: 4, 7, 10, 22, 43, 73, 139, 268 ($n = 3,\dots,10$) and m_n: 8, 14, 20, 44, 86, 146, 278, 536, 974, 1808 ($n = 1,\dots,10$) calculated in this way.

Also we can use the classical approach to derive a closed formula for m_n. Such a formula is connected with another formula of interest, namely that for f_n, the size of the female reproducing population at time $t = t_n$. Besides initial conditions (1) and recurrence relations (2) and (3), this approach involves a cubic equation and elementary properties of complex numbers. We observe from (3) that if f_n are proportional to the nth power of some real number a, then a should be a solution to the auxiliary equation:

$$(4) \qquad\qquad x^3 = x^2 + 3.$$

A standard approach shows that

$$(5) \qquad\qquad a = \frac{y^{1/3} + 1 + y^{-1/3}}{3} \approx 1.863707,$$

where $y = 41.5 + 4.5\sqrt{85}$. However the values of f_0, f_1, f_2 force us to represent f_n as a linear combination of the nth powers of all roots of equation (4): one real root a and two complex conjugates b and $\bar{b}$. It is not difficult to see that

$$(6) \qquad\qquad b = \frac{a - a^2 + i\sqrt{9a+3}}{2a}.$$

Since is f_n real, coefficients for b^n and $\bar{b}^n$ in our linear combination should be complex conjugate numbers. We have for all n,

(7)
$$f_n = \alpha a^n + \Re(\beta b^n),$$

where a real coefficient α and complex coefficient β should be set according to the initial conditions (1). Thus the following equations are satisfied:

(8)
$$\alpha + \Re(\beta) = \alpha a + \Re(\beta b) = \alpha a^2 + \Re(\beta b^2) = 1.$$

Equations (8) and (6) allow us to express coefficients α and β in terms of a. Then we use (7) to produce a formula for f_n in terms of a. To obtain a formula for m_n we use (1) and (2) to show that

(9)
$$m_n = 2 + 6(f_0 + f_1 + \cdots + f_{n-1}).$$

Equations (7) and (9) imply that

$$m_n = 2 + 6 \sum_{k=0}^{n-1} \left[\alpha a^k + \Re(\beta b^k) \right].$$

Since $a, b \neq 1$ we have

(10)
$$m_n = 2 + \left\{ 6\alpha(a^n - 1)/(a-1) + \Re\left[\beta(b^n - 1)/(b-1) \right] \right\}$$

for each n.

The exact formulas for f_n and m_n ($n = 0, 1, \dots$) in terms of a are given below:

(11)
$$f_n = \frac{1}{a^2 + 9} \left\{ a^{n+3} + 6 \cdot 2^{-n} \Re\left[\left(1 - ia/\sqrt{9a+3}\right)\left(1 - a + i\sqrt{9a+3}\ /a\right)^n \right] \right\}$$

and

(12)
$$m_n = \frac{2}{a^2 + 9} \left\{ a^{n+5} - 3(a-1)2^{-n} \Re\left[\left(a + 2 + \frac{i(9-2a)}{\sqrt{9a+3}}\right)\left(1 - a + \frac{i\sqrt{9a+3}}{a}\right)^n \right] \right\}.$$

These formulas and (5) can be used for calculations and further analysis of the sequences $\{m_n\}$ and $\{f_n\}$. For example (11), (12), and (5) imply that $m_n / m_{n-1} \approx f_n / f_{n-1} \approx a$ and $m_n = 7f_n$ for large n. Some approximate formulas in the trigonometric form are good for applications. Equations (5), (11), and (12) or alternatively, (5), (6), (8), (7), and (10) allow us to have approximations for f_n and m_n with any required degree of accuracy. For example, we have the following formulas:

$$f_n \approx 0.5189765 \cdot (1.863707)^n - 0.5215636 \cdot (1.268738)^n \cos(157.2605° + n109.9001°),$$

$$m_n \approx 3.605226 \cdot (1.863707)^n - 1.679117 \cdot (1.268738)^n \cos(17.06058° + n109.9001°).$$

which guarantee good accuracy for many values of n, in particular, to the hundredth for $n \leq 10$. A sophomore, Viet Bui, another student contributor majoring in biology, provided a detailed numerical analysis. For $t_n = 400$ days, these formulas give $f_{20} \approx 132,707$ and $m_{20} \approx 921,366$. The corresponding exact values are 132,706 and 921,362. The formula for m_n as well as some of its approximations can be used to solve both the time equation and the budget related question. In particular, one can use the formula $m_n \approx 3.605226 \cdot (1.863707)^n$ which is asymptotically good. This approximation shows that Brandon's intuition deserves full credit.

We note that the computational approach based only on recurrences will be of a key role if we need more sequence restrictions. For example, taking into account that reproducing females die in the end of their sixth reproduction period makes us replace equation (4) with an auxiliary equation of the seventh degree.

Having estimated the expenses, Brandon finds out that it costs \$6.40 to house one mouse over a 20-day period. So it costs $\$6.40 \cdot A_n$, where $A_n = \sum_{k=0}^{n-1} m_k$, to house the resulting population of mice over the $20n$-day period, $n \geq 1$ (assuming that most times the mice are kept separately). According to this result the cost for the 400-day period is about seven million dollars!

The mouse project belongs to a large class of projects involving population analysis. Another interesting problem of this type was considered in the project "Mathematical analysis of manatee population decrease." Bruce Ackerman, research scientist in Florida Marine Research Institute (Florida Fish and Wildlife Conservation Commission, St. Petersburg, Florida), proposed this problem. He has been conducting a long-term study of manatee (*Trichechus manatus latirostris*) mortality in Florida since 1974 (cf. [1]). Zocaris Vega, a junior biology major from my life sciences calculus class, was a student contributor to the project. Having employed the data on manatee mortality she used some statistical analysis to get the probability of their survival.

Conclusion

In conclusion I feel that the following points are the core of this paper:

- collegiate mathematics education for non-mathematics majors should be connected with relevant and practical community problems;

- a good deal of problems brought up by specialists in various areas inspire students and create service-learning opportunities for them;

- recognition of the above issues can improve the quality of mathematics education and make the role of mathematicians more visible in our communities.

References

1. B. B. Ackerman, S. D. Wright, R. K. Bonde, D. K. Odell, and D. J. Banowetz, "Trends and patterns in manatee mortality in Florida, 1974–1992," in *Population Biology of the Florida Manatee (Trichechus Manatus Latirostris)*, T.J.O'Shea, B. B. Ackerman, and H.F. Percival, eds., National Biological Service, Information and Technology, Report 1, 1995.

2. Committee on the Undergraduate Program in Mathematics, *Recommendations for a General Mathematical Sciences Program*, Mathematical Association of America, Washington, DC, 1981.

3. Committee on the Undergraduate Program in Mathematics, *Undergraduate Programs and Courses in the Mathematical Sciences, CUPM Curriculum Guide*, Mathematical Association of America, Washington, DC, 2004, available on-line: www.maa.org/cupm/.

4. M. E. Fowler, ed., *Zoo and Wild Animal Medicine*, W. B. Saunders, Philadelphia, 1978.

5. Mathematics Umbrella Group at USF, available on-line: www.math.usf.edu/mug/.

6. Dave Morgan, R. Kennedy Keller, "What evidence would prove the amyloid hypothesis? Towards rational drug treatments for Alzheimer's disease," *Journal of Alzheimer's Disease*, 4 (2002), 257–260.

7. J. Tan, T. Town, F. Crawford, T. Mori, A. DelleDonne, R. Crescentini, D. Obregon, R. A. Flavell, and M. J. Mullan, "Role of CD40 ligand in amyloidosis in transgenic Alzheimer's mice," *Nature Neuroscience*, 5 (2002), No. 12, 1288–1293.

About the Author

Arcadii Z. Grinshpan was born and educated in St. Petersburg (Leningrad), Russia. Since his student years at St. Petersburg (Leningrad) University he has been active in complex analysis and applied mathematics. He is the Director of Industrial Mathematics and Coordinator of the Mathematics Umbrella Group at the University of South Florida.

2.6

Designing Efficient Snow Plow Routes:
A Service-Learning Project

Peh H. Ng
University of Minnesota, Morris

Abstract. We describe our experience incorporating service-learning into a combinatorics and graph theory course. In particular, we present a discrete mathematical model which we used to design efficient snow plow routes for Morris, Minnesota. We also discuss several issues of service-learning within the course as well as beyond the classroom.

Introduction

In 1996, three mathematics faculty members of the University of Minnesota, Morris were awarded a SEAMS (Science, Engineering, Architecture, Mathematics, Computer Science) Grant from the Minnesota Campus Compact to incorporate service-learning into a mathematics course and two statistics courses. Part of the grant included the support of a coordinator, Mr. Ben Winchester, who took care of all administrative tasks ranging from setting up meetings between the students and the city officials to handling paper work involving expenditures incurred and surveys for assessment purposes. One reason that this service-learning project went smoothly in the course is the accommodating help of the coordinator.

Our objective in this paper is to discuss one of the projects that was done in the math course, Discrete & Combinatorial Mathematics, as well as touch upon a few issues to think about when incorporating service-learning into mathematics. This math course carries 4 credits, has about 12 students and, under the quarter and semester systems it lasts for 10 and 15 weeks, respectively. The course objectives are to help students:
- to learn the concepts of logic, proofs, and the language of mathematics;
- to understand the notions of discrete functions, sets, counting processes, and combinatorics;
- to learn the concepts of graph theory;

- to be aware of the connections between discrete mathematics and other math-related courses; and
- to be aware of the plethora of applications of discrete and combinatorial mathematics within and beyond the scope of academia.

The contents of this course include topics found in most discrete mathematics text books and topics in classical combinatorics and graph theory with applications.

It is mandatory for this course to have a course project where students find a real-world application of the material covered, write a paper on it and give an oral presentation to the class at the end of the course. When we incorporated service-learning, the main difference was that the particular projects were chosen to be of value to various community agencies within the city or county.

In the first section below, we describe a particular project which was of interest and beneficial value to the City of Morris, a small community of about 5600 people located in rural west central Minnesota. We also give the detailed mathematical model we used to solve a specific problem posed by the city. One of the objectives of the detailed description of the model is to illustrate the fact that in a process which is rich in both useful and useless data, it behooves all parties concerned to maximize their efforts at communications. In the next section, we summarize how we conducted the course, including how students worked on the projects, how they were graded, and how they communicated with the community agencies. In the last section, we discuss the rewards we reaped as well as challenges that we encountered throughout the process.

A Mathematical Model

There were several questions of interest that came from various community agencies, such as: What are the traffic-flow patterns around the city and its vicinity? How efficiently can we plow the streets in the city right after a snow storm? How do we group the city and its vicinity into several zones so that traffic crossing between zones is minimized? There is no room in this paper to describe all the models used to answer all the questions, but in this section, we will discuss one specific problem in detail so that you will have a rough idea of how elaborate the analysis could be based on just one specific question. At the end of this section, we list a few other common community-related projects that can also be modeled in a similar way.

Description of Project

We were asked to determine and design a cost-effective way to plow the streets and alleys so that each of the different types of snow plows and sanding trucks could traverse its group of streets and return to its depot. Pertinent information and constraints had to be considered. These include the following constraints: During a snow emergency period, there are a few roads that must be cleared first before others; costs are assumed to be proportional to the distance traveled by the snow plows; also, the city has five different classifications of plows (motor grater, reversible plow on a truck, one-way plow on a truck, a V-plow with a sander, a loader), and it has two motor graters and only one of each of the other types. Table 2.6.1 and the map in Figure 2.6.1 give a summary of the aforementioned information and constraints.

Table 2.6.1. Classifications of plows and the regions of the city

Plow type	Region Plowed
Motor grater (1)	Roads on the south-west of the city
Motor grater (2)	Roads on the north-east of the city
Reversible	Roads in south-east part of the city
One-way	Alleys in south-west of the city
V-plow	Alleys in north-east of the city
Loader	Parking lots and airport

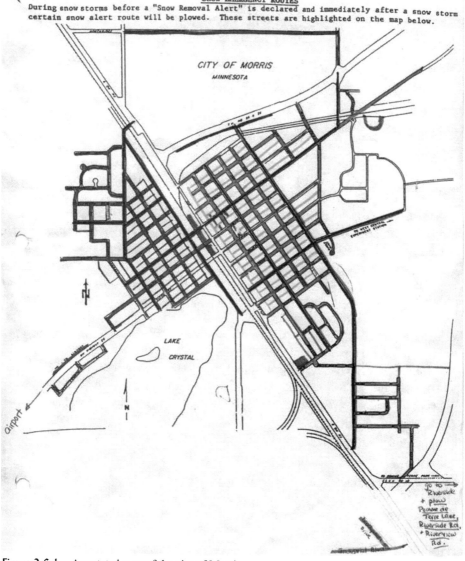

SNOW EMERGENCY ROUTES

During snow storms before a "Snow Removal Alert" is declared and immediately after a snow storm certain snow alert route will be plowed. These streets are highlighted on the map below.

Figure 2.6.1. Annotated map of the city of Morris.

A Bit of Graph Theory

The main concepts from discrete and combinatorial mathematics that were used to solve this problem come from graph theory and applications. To make this paper self-contained, we will give a brief introduction to graph theory and touch upon terminology that pertains to this project. For an in-depth coverage of graph theory, good references are Bondy and Murty [2] and West [3].

A *graph*, denoted, $G = (V, E)$, is a pair (V, E) where V is a set of *vertices* and E is a set of two-element subsets of V called *edges* ($E \subseteq \{(i,j): i,j \in V\}$). We will now define a few structural terms in a graph, $G = (V, E)$. Refer to Figure 2.6.2 for an example.

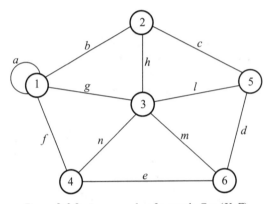

Figure 2.6.2. An example of a graph $G = (V, E)$.

For $v \in V$, an edge of the type (v,v) is called a *loop*. If $u, v \in V$ where $u \neq v$ then (u,v) and (u,v) are called *parallel edges*. The *degree* of a vertex is the number of edges incident to the vertex, with a loop counting twice. A graph G is *simple* if it contains no loops and no parallel edges. A *walk from vertex v_0 to vertex v_k* is a finite sequence $v_0, e_1, v_1, e_2, v_2, \ldots, e_k, v_k$, where each $v_i \in V$ for $i = 0, 1, 2, \ldots, k$ and each $e_i \equiv (v_{i-1}, v_i) \in E$ for $i = 1, 2, \ldots, k$. A *trail from vertex v_0 to vertex v_k* is a walk from vertex v_0 to vertex v_k which contains no repeated edges, while a *path from vertex v_0 to vertex v_k* is a trail from vertex v_0 to vertex v_k which contains no repeated vertices. Examples of a trail and a path from vertex 4 to vertex 1 in Figure 2.6.2 are $4, f, 1, a, 1, b, 2, h, 3, l, 5, d, 6, m, 3, g, 1$ and $4, e, 6, m, 3, g, 1$, respectively. A *tour* is a trail from vertex v_0 to vertex v_0 and a *cycle* is a trail from vertex v_0 to vertex v_0 which contains no other repeated vertices. A graph $G = (V, E)$ is said to be *connected* if there exists a path between any two vertices $u, v \in V$.

Optimal Eulerization of a Graph

We will now describe how to model the *efficient snow plow routes design* as an optimization problem on a graph. First, based on each snow plow's designation area on the map in Figure 2.6.1, we construct a graph $G = (V, E)$ by creating a vertex for each possible intersection. An edge is added between two vertices (intersections) if the block between the two intersections is to be plowed. For instance, the road system in Figure 2.6.3 translates to the graph in Figure 2.6.4 with appropriate costs, which are represented by the length of the particular block.

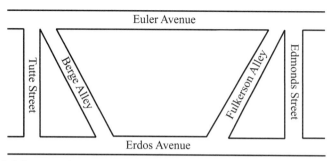

Figure 2.6.3. A snapshot of a road system (not drawn to scale).

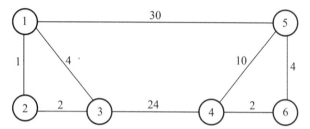

Figure 2.6.4. The graph that represents the snapshot of a road system in Figure 2.6.3, with numbers on edges corresponding to costs of traversing the edge.

Once the graph is constructed with the appropriate costs indicated on the edges, solving the city's problem is equivalent to finding a way to traverse each edge of the graph at least once, starting and returning to the same vertex, and in such a way that the total distance traversed is minimum, i.e., with as few repetitions on the edges as possible. In graph theory language, this translates to finding the optimal way to eulerize a graph if the graph is not *eulerian*. (By definition, a graph G is *eulerian* if it has an euler tour, meaning there exists a tour on G that visits all edges of G.) We found the following results useful:

Result 1. (Characterization of eulerian graphs.) *A graph $G = (V, E)$ is eulerian if and only if it is connected and all vertices have even degrees.*

Result 2. *In any graph $G = (V, E)$, there is always an even number of vertices with odd degrees.*

It is not difficult to see why Result 2 is true: when we sum the degrees of all the vertices in G we get twice the number of edges, i.e., $\sum_{v \in V} d_G(v) = 2|E|$. Thus, there must be an even number of odd terms on the left of the sum.

By and large, graphs obtained from maps of certain parts of cities are not eulerian; they usually are connected but they violate the even degree requirement. Thus, we need to find an efficient way to repeat edges so that the new graph is eulerian, i.e., *optimal eulerization of a graph*. Even after the graph has been eulerized, or if by chance the given graph is eulerian, we still need to construct an euler tour on the graph. In summary, given a connected graph $G = (V, E)$ that represents the road map and costs c_{ij} for all $(i, j) \in E$, we first find an optimal way to duplicate edges in G so that it is eulerian, and then construct an euler tour.

The algorithm in Figure 2.6.5 gives an optimal method to eulerize a connected graph if it is not eulerian. As an illustration of the algorithm in Figure 2.6.5, refer to the graph

Given: A connected $G = (V, E)$.

Want: A set of edges from E to duplicate so that the new graph $G' = (V, E')$ is eulerian.

Step A   Check to see if G is eulerian by checking if G is both connected and all its vertices have even degrees. If it is eulerian, then we are done. Otherwise, there exists at least one vertex of odd degree.

Step B   By Result 2, there is an even number of odd degree vertices. Thus, apply the Floyd-Warshall's algorithm to find the shortest path on G between every pair of the odd degree vertices. (For details of the algorithm, see Ahuja, Magnanti and Orlin [1]).

Step C   Define a new graph $\bar{G} = (\bar{V}, \bar{E})$ where $\bar{V}$ is the set of vertices in V with odd degrees, and $\bar{E} = \{(i, j) : i, j \in \bar{V}\}$. And, for each $(i, j) \in \bar{E}$, assign weights, $w_{ij} = $ length of the shortest path between vertices i and j from Step B; think of w_{ij} as the weight of grouping vertices i and j as

Find a *perfect matching* of $\bar{G}$ with minimum total weight, i.e., find a way to pair up the vertices in $\bar{V}$ such that the total of the pair-up weights is minimum. Since the number of vertices in $\bar{V}$ is even, we can always find a perfect matching. (See Ahuja, Magnanti and Orlin [1] for details of finding optimal perfect matching.)

Step D   With the perfect matching from Step C, we duplicate edges in G in the following way: Suppose vertex u and w are paired up from Step C, we duplicate each edge on the shortest path between vertices u and w obtained in Step B. Do this for every pair of vertices in the perfect matching.

Step E   Let E^d be the set of edges that were duplicated from Step D, and let $E' = E \cup E^d$. Then the new graph $G' = (V, E')$ is eulerian.

Figure 2.6.5. Algorithm for eulerizing a connected graph.

in Figure 2.6.4, which is connected but not eulerian; vertices in the set $\{1,3,4,5\}$ are of odd degrees.

In Step B, a summary of the results of Floyd-Warshall's algorithm is given in Table 2.6.2. For Step C, since we only have four vertices of odd degrees and since we only have three possible matchings, namely, either $\{(1,3), (4,5)\}$ or $\{(1,4), (3,5)\}$ or $\{(1,5), (3,4)\}$, the one that yields the minimum weight is $\{(1,3), (4,5)\}$ with total weight 9. For Steps D

Table 2.6.2. Results of shortest paths between odd degree vertices.

i, j	Shortest path between vertices i and j	Length of the shortest path
1, 3	$1 \leftrightarrow 2 \leftrightarrow 3$	3
1, 4	$1 \leftrightarrow 2 \leftrightarrow 3 \leftrightarrow 4$	27
1, 5	$1 \leftrightarrow 5$	30
3, 4	$3 \leftrightarrow 4$	24
3, 5	$3 \leftrightarrow 4 \leftrightarrow 6 \leftrightarrow 5$	30
4, 5	$4 \leftrightarrow 6 \leftrightarrow 5$	6

and E, the duplicated edges are those on the shortest paths between 1 and 3 and between 4 and 5, namely, $E^d = \{(1,2), (2,3), (4,6), (6,5)\}$. Clearly, after adding edges of E^d to G, the new graph, G', is eulerian.

To construct an euler tour given a graph that is eulerian, refer to any efficient algorithm based on the contructive proof of Result 1. (See Ahuja et.al. [1]). Finally, it is the euler tour constructed from G' that is interpreted as the efficient route for the snow plows.

Although the problem of designing efficient snow plow routes was posed by the public works department of the City of Morris and was a one-time project, the specific mathematical model and techniques used above are directly applicable to other problems that can also be modeled as finding an optimal eulerization of a graph. In particular, finding efficient ways to collect garbage on every block of the city, sweep the roads, do maintenance checks on an area with many railroad systems, and of course, deliver mail to all houses. Indeed, the above algorithm for finding an optimal way to eulerize a graph was proposed to solve the so-called Chinese Postman's Problem. (See Ahuja et al. [1]). One word of caution is that there are insidious consequences if one mistakenly models a real-world problem as an optimal routing problem on visiting vertices instead of one on visiting edges. The latter is efficiently solved by the technique described in this section while the former has no known efficient method to solve and probably never will.

We would also like to mention that the mathematical process, in the pedagogical sense, is similar to the description in this section even if the mathematical model used is different. For instance, in one of the posed questions on how to group the city and its vicinity into several zones so that traffic crossing between zones is minimized, once we know the number of zones, say k, then a mathematical model to use would be finding a minimum k-cut on a graph that is constructed in a similar way to the one that was created for the snow plow routes.

Administrative Process Throughout the Term

In the introductory section of this paper, we gave a description of the course and its contents. Now, we will summarize the administrative process of conducting the course, including how students worked on the projects, how they were graded, and how they communicated with the community agencies.

Prior To and As The Term Starts

Several questions of interest came from various community agencies. Before the course started, the instructor (author) spent a couple of weeks grouping the questions into two categories, namely, transportation and network flows, and network designs. The question, "what are the traffic-flow patterns around the city and its vicinity?", would fall into the first category. "How efficiently can we plow the streets in the city right after a snow storm?" would fall into the second category. "How do we group the city and its vicinity into several zones so that traffic crossing between zones is minimized?" would fall into both categories. When the course started, we divided the students randomly into three groups, each of which were randomly assigned projects within the categories mentioned above. For instance, we assigned the project described in Section 2 on designing efficient snow plow routes to a team of four students.

A course project is mandatory when we teach this course, and the project carries the same weight as one of the three within-term exams. The requirement of the course project is for students to discuss and analyze an application of a topic covered in the course; they may have to look ahead in the syllabus to browse the topics. Students usually work on these projects in pairs; each group is required to submit a written report and give an in-class 20-minute oral presentation at the end of the term. When we incorporated service-learning, we chose to do projects that were of value to the community in which we were a part.

As the Course Progressed

Each student group met regularly outside of class time to work on their project over the course of the entire term. It is advisable to have a group leader whose main job is to coordinate these meetings and keep a central working document of the written report.

The value of communication among all parties involved, i.e., the students, the instructor, the service-learning coordinator, and the community agencies, should not be underestimated. We (instructor and student groups) met on a weekly basis outside of class time and by appointment to discuss issues of, and progress on, the projects; the length of the meeting was very short at the beginning of the term and was much longer (about an hour) towards the end of the term. The instructor was a facilitator in making sure that timely progress occurred and that the analyses done and reports written were mathematically correct.

The coordinator of our SEAMS Grant was the liaison between the community agencies and the class (students and instructor). We cannot put enough emphasis on how important it is to have a liaison who is familiar with the people within the community as well as the instructor and students, and who has some knowledge of the contents of the course. Whenever we had specific questions on a piece of data or information from the agencies, the coordinator came back with responses within a couple of days or amicably pressured the community agencies for prompt responses. This is in addition to several short meetings that the coordinator set up during the first third of the course. There was an instance during the term when students felt that a face-to-face discussion on several small issues that arose in the middle of their project was necessary with a particular agency. The coordinator contacted the people of that agency and immediately scheduled a short visit to class within a few days.

At the End of the Term

Within their groups of about 4 persons each, the students did all the analyses on their own, with their instructor as an advisor and expert to bounce technical questions off. Also, they wrote their final reports and gave oral presentations to which people from Campus Compact and the community agencies were invited. Again, their instructor advised the students on the organization of their written reports and gave feedback on their trial presentations.

In terms of evaluations, a university-required course evaluation and a post service-learning evaluation from Campus Compact were given to the students at the end of the term. The only written "reflection process" on the students' part came from what they said in these evaluations.

As for grading the project, which carries about 15% of the course grade, half of the project grade is on the final report and the other half on the in-class oral presentation. The instructor was the only person who graded written reports, while the oral presentations were evaluated by all who attended. Attributes used on the evaluations of the written and oral presentations included how well students communicated the mathematical ideas, how well they demonstrated their content knowledge of their research project, the level of independent and critical thinking skills, clarity and organization, and enthusiasm. Each student within a group received the same grade on the project.

The Good News and the Challenges

In this section, we will briefly discuss a few rewards and challenges that we faced during the semester when we incorporated service-learning into Discrete and Combinatorial Mathematics. It is common for us to run into a few obstacles whenever we try to add a new aspect into our pedagogical styles, which we are very familiar and comfortable with. Since the idea of incorporating service-learning into our curriculum was a rather unfamiliar concept especially in the mathematical sciences, it was only natural that this was a challenging experience with a rather steep learning curve. Notwithstanding the challenges, we found the outcomes of the project to be beneficial to the students, instructor, the university and the community.

The Good News

A former Vice-Chancellor for Academic Affairs and Dean of the University of Minnesota, Morris once said, "The math project on designing efficient snow-plow routes for the city is a poster child for service-learning projects." This project has a strong academic component as well as a service component.

From an academician's view point, we found that doing service-learning projects has benefits that cover our mantras of excelling in teaching, research and service. In terms of teaching, a hands-on and real-world application of the mathematical theory covered in class provides a motivational tool for the instructor. Being able to work on a modeling problem in which you have direct contact with all the raw data and the people using it, is definitely refreshing, albeit challenging; thus it complements working on relatively simple textbook problems. While the process of solving the problem is going on, we noted additional research questions such as "what model would fit with extra constraints, i.e., what if there must be sequential execution of different machines?" and "can we decompose the graph into simple cycles so that we are guaranteed to cover each block in two of the cycles?" The first question is a modeling one while the second is more of a basic research question in graph theory. Last but not least, this project clearly is of service or outreach value to the community we reside in.

From the students' perspectives (which were obtained via written student evaluations at the end of the course), in addition to the mathematical content learned, they felt a sense of gratification and empowerment, knowing that what they did actually made a difference to the community they lived in or will live in for at least four years. Students were able to get a taste of being in the role of applied mathematicians and as valuable members of a research team. In addition, students had first hand experience in dealing and communicating with non-mathematicians and non-academicians; they learned what questions to ask and how to identify useful information that is pertinent to their project. Indeed, a few years after their graduation, students gave us unsolicited comments on how they used this experience in their careers, where they often work on group projects and as professional consultants.

In terms of the benefits to the university and the community, service-learning projects like this one provide opportunities to bring together some of the needs/problems of the community and the resources from the university campus. For a small town, usually there is limited funding to hire external organizations to do extensive analyses or problem solving. And, at a public institution of higher learning, the chance to apply theoretical ideas beyond the walls of the classroom is invaluable, let alone applying it directly to the surrounding community. It was a win-win situation for both parties.

Last but not least, these projects encouraged future partnerships and collaborations with groups beyond the university. In fact, after the end of the SEAMS Grant for the three mathematics and statistics courses at the University of Minnesota, Morris, several other courses and projects with service-learning components were conducted successfully by faculty across the campus in conjunction with different organizations in the city.

The Challenges To Watch For

In this section, we will discuss a few challenges that we faced and how we dealt with these challenges while working on this service-learning project.

Time Constraints

Whether the duration of the term is a 10 week quarter or a 15 week semester, the length of time is quite restrictive from the students' point of view in terms of trying to learn the material and then applying the concepts to their specific projects. We took on the service-learning project with the understanding that it would not require us to compromise on the contents of the curriculum, but then we had to teach the students the theoretical concepts before they could apply them. Timing within the course was especially crucial for the group of students who worked on the efficient snow plow routes project; we did not finish covering the theory of optimal eulerization of a graph, optimal perfect matching, and shortest paths until about two thirds into the course.

This does not mean that students did not start on their projects until they learned the specific mathematical concepts; a great deal of preliminary work was done on the information given to the class. As we mentioned at the beginning of this paper, having a full-time coordinator who was a wonderful liaison between the class (students and instructor) and the community agency was invaluable. The coordinator was able to schedule numerous meetings outside class, and even brought the director of public works into two class sessions. A few meetings with representatives from the community agencies took place during the first third of the course, so students had a good "big picture" idea of their ultimate goal on their specific project.

Resources and Communications

Incorporating service-learning requires time and resources. Unless you (faculty) have a lot of time to spare, our advice is that you apply for some grants (external or internal) so that you can afford to have a person who coordinates laborious administrative tasks and paper work. In addition, it is very important to have some discretionary funds when student groups need to make a visit to the community agency or to go on a field trip that is essential to their project. It would have been impossible for us to implement service-learning had we not received such external funding. One good place to start searching for funding opportunities is Campus Compact, a national coalition of about 850 college and university presidents committed to the civic purposes of higher education.

Another challenge throughout all of these service-learning projects has been keeping up with all of the communications with community organizations. Delays are inevitable; there could be complex bureaucratic channels in an agency through which we might need to obtain some pertinent data or information. We were very fortunate to have an outstanding coordinator who had enough intimate knowledge of the community organizations to minimize delays and to promptly schedule meetings with the people concerned.

Process of Modeling

Not surprisingly, we found that pruning the questions given to us and fitting them into a mathematical context were daunting and time-consuming tasks. This issue is not restricted to working on community projects; this is an accepted occurrence among professional applied mathematicians who work on problems from industry. On the other hand, the process of going from plain questions to mathematical models is the fun part of being a practicing applied mathematician.

To minimize misunderstandings and complications from the questions posed by the community agencies, in the case of service-learning within a course with a specific starting time and a set length of time, our advice is for the instructor and coordinator to carefully plan their first organizational meeting with the students. In particular, after initial contacts and discussions with the community organizations, we advise the instructor to do some preliminary pruning of the questions before the course starts. The fact of the matter is that community organizations might have restrictions and deadlines that do not coincide neatly with any academic calendars.

Follow-ups

Even though the project was completed by the end of the term, we (instructor and coordinator) did some follow-ups with the community agencies. We strongly advise you to do something similar to this either for the purpose of closure or for the sake of future collaborations. In our case, we were invited to give presentations of our students' results at a regularly scheduled meeting of the city and its planning commission, so that the city had the opportunity to ask questions and make comments on any proposals that students made. Each of the community agencies was also given copies of the final reports written by the students. One reality we observed was that it takes time to implement changes, especially at state-run agencies. Nevertheless, changes do occur slowly but surely.

Closing Remarks

Although the project we discussed in this paper was done in a mathematics course, we would like to make it clear that the contents of discrete and combinatorial mathematics are very interdisciplinary. If you have a course in computer science or engineering that covers similar topics, and if you are proactively seeking service-learning projects, you can approach your community to see if this is applicable.

Overall, the integration of service-learning into discrete and combinatorial mathematics was arguably a success, notwithstanding the obstacles that the students encountered. The results and proposals recommended by the students were very thorough, meticulously thought out, and well-organized. This experience was the beginning of a collaborative partnership between the City of Morris and the University of Minnesota, Morris.

References

1. Ahuja, R.K., T.L. Magnanti and J.B. Orlin [1993], *Network Flows: Theory, Algorithms, and Applications*, Prentice Hall.
2. Bondy, J.A. and U.S.R. Murty [1976], *Graph Theory with Applications*, North-Holland.
3. West, D.B. [2001], *Introduction to Graph Theory* (second edition), Prentice Hall.

About the Author

Peh Ng received a BS in Mathematics and Physics from Adrian College, Michigan; an MS in applied mathematics from Purdue University; and a PhD in Operations Research and Combinatorial Optimization, also from Purdue University. She is currently a Morse-Alumni Distinguished Teaching Professor of Mathematics and an Associate Professor of Mathematics at the University of Minnesota, Morris. Her areas of research include operations research, discrete optimization and graph theory, within which she has several publications.

Chapter 3

Service-Learning in Statistics

3.1

Perspectives on Statistics Projects in a Service-Learning Framework

Gina Reed
Gainesville College

Introduction

In 1992, significant efforts towards undergraduate statistics education reform began to emerge. The Mathematical Association of America (MAA) published *Statistics for the Twenty-First Century, Perspectives on Contemporary Statistics* [2], and *Heeding the Call for Change* [8]. In these volumes, many statisticians agreed that there was a gap between statistics teaching and statistical practice. The American Statistical Association (ASA)/ Mathematical Association of America Joint Curriculum Committee made three recommendations to improve the teaching of statistics: emphasize the elements of statistical thinking, incorporate more data and concepts, and foster active learning. Instruction in modern statistics must begin with data analysis, both because concrete experience with data motivates the more abstract parts of our subject and because exploring even haphazardly produced data can provide insight [3].

Since those first discussions began, many professors have experimented with curriculum reform in the statistics classroom. One of the most exciting possibilities that incorporates the recommendations of the ASA/MAA Joint Curriculum Committee is service-learning. Service-learning is a concrete application of statistical methods using real data and leading to analysis and interpretation that is useful to a community organization. Service-learning projects emphasize the elements of statistical thinking because using real data gives students the opportunity to explore the statistical concepts presented in class and receive practical experience at the same time. Service-learning incorporates more data and concepts in the course because it is a bridge between the classroom and the real world. Many organizations and agencies produce data that need to be analyzed and interpreted in order to be of use to the group. Service-learning fosters active learning because the students must take what they have heard in class and learn how to apply it to the service-learning activities. Thus service-learning can reinforce and deepen the learning objectives in statistics courses.

Moore [4] argues that the most effective learning takes place when *content* (what we want them to learn), *pedagogy* (what we do to help them learn), and *technology* reinforce each other in a balanced manner. The four essays that follow illustrate in detail several types of service-learning projects. The ideas in the essays can be a springboard for you to create service-learning projects appropriate to your own statistics courses.

The Scope of Service-Learning in Statistics

One of the most exciting things about service-learning in statistics is that it can be used at any level of coursework, ranging from very elementary courses emphasizing descriptive statistics to quite advanced courses in methods of inference. The sections that follow illustrate this range. My own experience with service-learning has been both at the introductory level and at the honors level. Service-learning projects work equally well in both courses with the amount of instructor guidance varying based upon the abilities of the students and the needs of the client agency.

There are several textbooks that are well suited to a class with a service-learning component. The most common text mentioned in the essays is *Introduction to the Practice of Statistics* by Moore and McCabe [5]. Others that adopt an active-learning approach include *Workshop Statistics* by Rossman and Chance [6] and *Activity-Based Statistics* by Shaeffer, et al [7]. In addition, many faculty develop supplemental computer laboratories and/or classroom activities that closely mimic some of the statistical procedures that the students will be expected to utilize in the data analysis section of the service-learning report.

The kinds of institutions that use service-learning successfully are also quite varied. Service-learning projects can be found in high schools, community and two-year colleges, and private and public four-year colleges and university. The essays include a mixture of experiences from all types and sizes of institutions of higher education.

One of the largest hurdles to overcome when implementing service-learning projects is to make contacts with potential clients, but the good news is that the pool is very large. Client categories include community agencies such as the American Red Cross, homeless shelters, and other non-profit agencies. Other sources of clients are health-related organizations, school systems, public libraries, and many others. If you look anywhere for organizations that rely on public support and that probably run on an austere budget, then you will also find organizations likely to welcome contributed services from statistics professors and their classes.

Some projects involve the students in finding agencies that are of particular interest to them, whereas in other settings they may be selected in advance by the faculty member. Whenever the student is allowed to choose the agency, the faculty member needs to look at the data and the research questions in order to determine if the project is within the parameters of the learning objectives of the class. Ultimately, the faculty member should reject or help to modify the proposed project if it will not be suitable. If the faculty member is choosing the agency, the initial contact can be by phone or letter. A sample of such a letter is given in the essay by Hydorn, Section 3.4. When I began my own service-learning activities, I contacted non-profit United Way agencies by phone and inquired if they were interested in free statistical consulting. If they answered affirmatively, I set up a meeting with the director and then visited to discuss their needs. Based on such a meeting, the faculty member can determine what needs the agency has and then decide if this project would complement the educational goals of the course. If the agency's needs and the course's objectives are complementary, a representative of the agency can then come to campus and speak to the class directly.

In some cases, the organization or agency needs help in preparing a survey to collect data. In other cases, they already have data but have not analyzed it. Since survey development, data collection and analysis are difficult to complete in the time span of a single course, there may be times when it is best to divide the project into two parts and have sequential classes complete the individual segments.

Diverse Faculty Approaches

The materials that have been developed by faculty also vary widely. This variation is based upon the individual needs of the agency and the educational objectives of the course. In some cases, design issues are paramount and in other cases, the agency has the data and wants specific questions to be addressed. In this latter scenario, the faculty member can outline the necessary procedures for the students or allow the students the opportunity to choose the appropriate statistical tools. Such faculty guidance can be accomplished in class, or with handouts, or by using web pages with similar projects available for student perusal. Other faculty choose to give simulated homework and/or lab assignments with specific instructions to "train" the students how to approach the service learning project. There are many different approaches, but all faculty agree there must be clearly defined guidelines for the students. One of the essays (Section 3.3) discusses a workbook that was developed by their faculty for student use. Another (Section 3.2) uses a "writing associate" to assist students in their writing.

A variety of project assignments are discussed in the essays. Some projects are optional, leading to extra credit or substituting for a test or other course requirement, while others are mandatory. They can be short-term or long-term, with the latter being more common. Usually, these more time-consuming projects count as a higher portion of the final grade. Most faculty require written reports that contain certain components such as an executive summary, an introduction to the agency and their requirements, a survey description, data description with a list of possible biases, the analyses including graphs and tables, and the conclusions. For instance, I count the project as 20% of the student's final grade because I expect a final written report with all of the above components as well as a comparison of this years' data to last years' data, when appropriate. In my honors class, there must also be an oral presentation on the project, something that is fairly common in most service learning projects when time permits. I recommend an early draft of the written report being due at least three weeks before the end of the semester in order to allow students to incorporate the faculty member's feedback into the final written report. This ensures the accuracy and quality of the findings before the student's oral presentation to the agency. Without feedback from the instructor, the quality of some reports can be poor. Also, faculty should consider making adjustments in the course's traditional workload because of the additional work often associated with the service-learning project.

The assignment of the project can be configured in different ways. Students can work individually or in groups. The whole class can be working on a single project that is subdivided into small parts and will be compiled for the agency at the end of the term or the students may be working for different agencies, thus leading to many different, unrelated final reports. The decision should be based on the amount of work the agencies need. Whether the college is residential or a commuter institution will help to determine if group or individual work is preferable. In my introductory class, I select one agency and assign groups to work on portions of the total project. Then I compile the final reports into a single report. In the honors class, the whole class works together and produces a single written report. Then the class gives an oral presentation to the agency.

If groups are used, it can be very valuable to obtain an assessment of each group member by the other group participants. This assessment process needs to be discussed in class prior to the group's initial meetings. The faculty member needs to take an active role in explaining what is and what is not an acceptable division of labor. The individual member evaluation needs to be recognized as a component of the final grade that each student will receive in addition to the grade on the group report. These precautions ensure more equal distribution of work and fewer difficulties between group members as the semester progresses. Of course, this is not necessary if the students work individually. Other assessment tools are pre- and post-tests, supervisor reports from the agency, and exam questions. I also find it helpful to meet with the agency after the semester is over to clarify any questions or concerns that may have arisen and were not addressed adequately during the semester.

The technology needed for service-learning projects include a word processing package such as Word and adequate statistical computing software such as SAS, SPSS, MiniTab, WebStat, or Excel, depending on the level needed. Also, PowerPoint is very useful if there is an oral presentation component for the class or the agency partners. The preferred technology should be used in classroom demonstrations so the students are familiar and comfortable in using the appropriate technology while working on the service-learning project.

All true service-learning must include not only some form of civic engagement but also a reflective process. Some suggested guidelines for the reflective activities are that it link the experience to the learning, occur on a regular basis, and lead to systematic feedback. The possible tools in the student reflective/reaction process are written assignments like synthesis papers or a semester-long journal, oral assignments like focus groups or individual interviews, and observations by the agency supervisor or faculty member. Usually the reflective component is included as part of the service-learning project grade.

All of the authors in this chapter have found student feedback to the service- learning project to be positive. Students generally express the view that the service-learning component enhances their understanding of the material taught in class and they feel more motivated to learn. In my classes, the perceived usefulness of technology received higher ratings in classes that had a service learning component. Many students also comment that they enjoyed doing statistics since they knew they were helping others in the process. There have been some complaints about working in groups. However, research by Garfield [1] indicates that allowing students to work collaboratively in class, solving problems and communicating their understanding, better prepares them to apply their learning in future courses and careers. Authors of the essays describe their assessment tools for student reaction to the projects. Several use pre- and post- questionnaires. Feedback from assessment tools allows for improvement in the projects each time service-learning is a component of another statistics course.

Conclusions

The benefits of using service-learning in statistics are many. The classroom environment becomes more experiential with active learning. The project provides richer examples for class discussion and the students more fully participate in these discussions. The students encounter the unexpected with real data that is not present in data sets in textbooks, which in turn provides them with opportunities to explore the application of ideas and concepts to new situations. There is more collaborative interaction between faculty, students and agency representatives. The students experience the power of statistics in a realistic manner and they feel a sense of accomplishment in helping others that is rarely attained in a traditional classroom. In fact, some of my students continue working with the agencies after the course is over. My experience is that the agencies are quite pleased with the results, and they continue to participate in service learning projects year after year. For faculty, this is a professional community service activity that can be noted on self evaluation reports. For the college, it is community outreach with positive public exposure.

There are some drawbacks to the process, however. The major one is the time commitment required of the supervising faculty member. Service-learning projects are not as easy to organize and keep on schedule as a traditional class because you are working with an outside agency that does not work on an academic calendar. Faculty members must be flexible and adaptable in order to handle the uncertainty and variation that happens along the way. However, several authors mention that they received grants which allowed for release time or the hiring of student assistants to equalize the time commitment.

I encourage interested readers to try service-learning in your own courses. Based on my experiences and those of the other authors, service-learning can revitalize a class as the students become more engaged in learning. Since I began using service-learning, I have not had students asking how they will ever use statistics. This is certainly a step in the right direction!

I would like to thank Allan Rossman for his encouragement and helpful input as I was preparing this essay and Charles Hadlock for providing me with the opportunity to contribute to this volume.

References

1. Joan Garfield, "How Students Learn Statistics," *International Statistical Review*, 63 (1995) 25–34.
2. Florence Gordon and Sheldon Gordon, eds., *Statistics for the Twenty-first Century*, MAA Notes, No. 26, Mathematical Association of America, Washington, DC, 1992.

3. David S. Moore, "Introduction: What Is Statistics?" in *Perspectives on Contemporary Statistics*, MAA Notes, No. 21, David Hoaglin and David Moore, eds., Mathematical Association of America, Washington, DC, 1992, pp. 1–17.
4. David S. Moore, "New Pedagogy and New Content: The Case of Statistics," *International Statistical Review*, 65 (1992) 123–165.
5. David S. Moore and George P. McCabe, *Introduction to the Practice of Statistics*, W. H. Freeman and Company, New York, 2002.
6. Allan J. Rossman and Beth Chance, *Workshop Statistics,* 2nd edition, Springer-Verlag, New York, 2001.
7. Richard Shaeffer, Mrudulla Gnanadesikan, Ann Watkins, and Jeffrey Witmer, *Activity-Based Statistics*, Springer-Verlag, New York, 1996.
8. Lynn Arthur Steen, ed., *Heeding the Call for Change: Suggestions for Curricular Action*, MAA Notes, No. 22, Mathematical Association of America, Washington, D.C., 1992.

About the Author

Gina Reed is a Professor of Mathematics at Gainesville College in Georgia, where she is the statistics coordinator. She has initiated service-learning projects as discussed in this section at Gainesville College.

3.2

Making Meaning, Applying Statistics

Rob Root[*], Trisha Thorme[†], and Char Gray[*]

Lafayette College and †Princeton University

Abstract. The authors have created, taught, revised, and supported an applied statistics course that incorporates community-based projects. The course requires reflection on the meaning of data in contexts richer than classroom presentation can allow, and so leads students to a deeper and more nuanced understanding of statistics. We describe the course and the project and writing assignments and reflect on service-learning and statistics pedagogy.

Introduction

Applied statistics classes for students not going on in mathematics often include a project designed to integrate statistical concepts and computations into practice [2, 4]. In some of these classes, students may choose to do a community-based or some other kind of real world project [5]. Skeptics sometimes feel that the interplay between classroom and community detracts from the learning of course content, but much of the service-learning literature, our own efforts included, demonstrates that students do indeed learn the material. For example, see [7, 8, 9], which use benchmarks from current research, and [1, 3, 10], which show how projects are structured to meet that goal

This chapter describes a course that the authors have created, taught, revised, and supported at Lafayette College since 1998. The course includes a semester project in which students typically complete an elementary statistical study for a local community-based organization. Because the study allows students opportunities to use statistics in rich contexts, the project is an important tool both for learning to apply statistics and for motivating concepts presented in the course. The study typically requires students to learn about some aspect of the community and reflect on the value of data analysis for a community-based organization and its mission. The result is a heightened civic awareness deriving naturally from completion of the statistical study. Reflection on the application of statistical ideas and processes in rich contexts offers students an opportunity for deeper and more nuanced understanding of the utility and limitations of statistics. Thus the community-based assignment complements rather than supplements the core material.

As we have expanded the scope of our learning goals and our understanding of how students learn, we have made significant changes over the past four years of using a community-based pedagogy, incorporating expectations that students reflect on experiences in context through journals and reports and that they engage civic institutions in order to increase their citizenship skills. The addition of these goals in no way detracts from the learning of statistics; in fact it increases the power of the course as a learning opportunity by integrating novel pedagogical objectives into the achievement of standard ones.

This essay begins with a description of course goals, focused on the project. A description of the project assignment itself follows, with the rationale for its structure and components. We reflect on the strengths and weaknesses of the project as a pedagogical tool. Brief descriptions of a few representative projects offer a sense of what is done, for whom, and how participants feel about the process and results. We end with issues raised by our practice and comments on the utility of service-learning pedagogy.

The Course and Its Goals

The course is in applied statistics, and its primary goal is to give students an introduction to using statistics. This requires familiarity with descriptive statistics: randomness, variability, distribution, association, and correlation. The course includes elements of study design, and it is essential that students gain a practical knowledge of inference. This material can be conveyed in the classroom, and we have found no better way to introduce these ideas.

While the course uses traditional classroom data and relies on a text that offers a wealth of real data (Moore and McCabe's *Introduction to the Practice of Statistics*), encountering real data also outside the relatively antiseptic setting of classroom and homework exercises offers unique advantages. The course is structured around the semester project, which incorporates all aspects of experiential learning, as Table 4.2.1 demonstrates. Classroom presentation is reinforced by immediate use of material in the project, and these activities naturally take students through the four basic activities that comprise experiential learning per Kolb [6]. This table demonstrates that the structure for a statistical study that calls on students' knowledge as it is developed in the classroom also leads students through the cycle of activities expected for a service-learning experience to develop civic awareness through reflection on actual experience.

These goals are less concrete than those the instructor might typically set for other classes. For example, in parallel sections of this course, projects of smaller scope are assigned, but students are required to apply inference involving the t-distribution. As we will see, the projects described here do not delimit the kind of inference to be done, and sometimes do not require inference at all. This seems to be characteristic of service-learning; it requires a willingness to forego carefully delimited pedagogical goals, but allows for compensating richer learning environments and heightened motivation for subject mastery. For this course, this trade-off has proved overwhelmingly beneficial for the majority of students.

Table 4.2.1. The schedule of the course shows how the classroom presentation feeds into the semester project. The experiential learning activities associated with the different phases of the project are also shown.

Course Schedule — 14 week Semester

Week	Classroom Presentation	Report	Project Activity	Experiential Dimension
1	Descriptive Statistics Distributions and Associations	0	Choose topic and group.	Concrete experience. Reflective observation.
2				
3				
4	Statistical Design	1	Learn about context. Define driving question.	Concrete experience. Reflective observation.
5				
6		2	Describe design of study.	Reflective observation. Abstract conceptualization.
7	Probability and Random Variables			
8				
9		3	Collect and describe data.	Active experimentation.
10	Inference using the normal distribution			
11				
12	Inference using the t and χ-squared distributions	Final	Analyze, interpret and present results.	Reflective observation. Abstract conceptualization.
13				
14				

Choosing a Project

Many students choose a partner/client out of a real desire to support and learn more about their community. Others make this choice because projects are ready for them to pursue, already vetted by the instructor. Many students prefer to find their own partners on campus. The issues available to study on campus seem more immediate and relevant to these students. These studies do not always serve to expose students to their position in the larger community, but they typically provide a suitable venue for a semester project in statistics, requiring comparable reflection on the importance of context. Thus these are acceptable vehicles for the statistical goals of the course, and the instructor accepts suitable project topics from on campus sources despite lack of opportunities for broader civic engagement.

The first assignment for the project is distributed on the second day of class. It presents an overview of the entire project and includes a form for a preliminary report. This report has two components: a listing of team members, contact information, and common free time. This confirms that the group agrees to work together and has done the minimal planning necessary for the group to function cohesively. We were surprised by the need to spell out that students share contact information and establish common free time for collaboration, but our explicit requirements (particularly that students have available two hours a week to work together) signal the importance of the project and the expected level of effort.

The second component is a proposed community partner, whom the instructor investigates at this point if he has not previously worked with them. If the work being requested is useful to the proposed partner and appropriate for the course and size of the group proposing, the proposed partner and study are accepted, usually with some clarifications for both partner and group. It rarely happens that the instructor rejects a partner, but in this case the group is responsible for choosing a new partner, and due dates are adjusted to compensate for the delayed start.

Finding a Driving Question

The first report in which students actually make progress toward completing the project occurs after they have met with their partners. This report and the next two take the form of memos addressed to the partner. Students must provide minutes of their meeting with the partner and summarize the goal of their project as a driving question. (For a definition and discussion of driving question, see [12, 13].) The report must also include an outline of the planned delegation of responsibilities among team members. This is less a firm commitment than an attempt to get students thinking about how to collaborate effectively and efficiently.

This report is the first that is reviewed with a Writing Associate (WA, described below), and so provides the first opportunity for journaled reflection. In the journal assignment, students are encouraged to consider carefully their client's situation and possible questions that might be answered statistically. As Table 4.2.1 shows, this occurs when students are just finishing the section of the course on descriptive statistics, and so they have an appreciation for basic ideas of center, spread, and shape of distributions; and direction, strength, and form of a relationship between two random variables. These ideas are enough to frame the most basic questions, that is to say ones they will be learning to answer during the course. Formulation of the driving question occurs at an ideal moment for students to apply their newfound knowledge of descriptive statistics to describing the world. Reflection on a community issue is the necessary foundation for employing their academic knowledge.

Study Design and Project Types

The second report is a description of the design of the study together with a description of the foreseeable value of the data. The projects can be grouped in four types:
1. Studies designed and analyzed by the students, who must also collect the data.
2. Studies whose design has been specified by the client, but the students are to collect and analyze the data.
3. Studies analyzing data previously collected and provided by the client.
4. Studies in which the client believes the data are available, and has indicated a means for obtaining it. The students are responsible for obtaining data, or summary statistics, and for the analysis.

Each type (see sidebars for examples) presents characteristic learning opportunities. The first type would seem to be most challenging, since students are responsible for all aspects of the study. However, the types are listed in order of remoteness of students from the actual process of collecting data. As that distance increases, so does the difficulty they face in ascertaining sources of bias in the data, and in transforming data into useful information. This generally compensates for increasing ease in accessing data.

In this second report, teams must describe how data were collected and assess their usefulness in answering the driving question. For projects of the first two types, emphasis is on the first of these tasks. For students carrying out data collection, the study design should assure that the data collected will address all issues at hand. Projects of the last two types require more focus on the second task. This report is due after students have studied the design of statistical studies, and it is their first opportunity to consider the implications of design considerations on their ability to draw conclusions from data. It is remarkable that they make this attempt without benefit of knowledge of inference as a structure for measuring strength of statistical evidence. This omission is actually useful, since students deal with questions of quality of data in isolation from issues of statistical power. One way this shows up is that sample sizes for studies are set by the size of the group and time needed to collect the data, rather than by the need to obtain a specified power. While incorporating the latter into design considerations would be valuable, we can't accomplish that at present.

The second reports for projects of the fourth type are unique. Students find data relating to some issue by contacting public offices, always including some local offices that require in-person visits, and generally some that maintain web sites. The real difficulty in these projects is less the quality of data than extricating data and interpreting it. For this kind of project, the second report typically lists sources that the group has identified and kinds of data that they have either obtained or

Type 1 Project
Pretesting a survey for graduating seniors
Lafayette College's strategic plan calls for assessing opinions of various campus constituencies on a variety of issues. Students worked with the Provost and Dean of Students to develop and test a survey instrument to collect opinions of graduating seniors about academic advising, technology and library resources, and on-campus entertainment/sports. Students wrote two sets of questions concerning these issues and administered them to a random sample of graduating seniors. Analyses revealed the most useful questions.

believe that they can eventually obtain. They must consider comparability of data, including means of transforming data into roughly commensurable forms, what further data would be required to accomplish this, and bias that might result. In sum, because this type of project typically isolates students from raw data, it has requirements distinct from the other types, and offers learning opportunities that are directed more toward manipulation and interpretation of statistics than their construction.

Dealing with Data

The third report is a descriptive presentation of the data/statistics the group has collected. This report is often short on words and long on graphs. The report is due once the team

Type 2 Project
Surveying blood center volunteers

The local blood center's new volunteer coordinator wanted to find out how to increase the satisfaction of volunteers. The students assisted in creating the survey and cover letter that went out to all volunteers and analyzed the data from responses. In response to the students' recommendations, the director rescheduled the awards banquet and ironed out difficulties between volunteers and paid personnel.

has collected all its raw data and put it in digital form. The process of entering data into a statistics application is in itself instructive for students. Although they have no difficulty in completing textbook exercises that ask them to categorize variables as quantitative or categorical, discrete or continuous, the same issues in the unfamiliar territory of digitizing responses to a questionnaire or data collected on application forms of an organization (for instance) seem newly challenging. The context in which lessons are learned is everything, and revisiting lessons that seem trivial in an academic setting now in the richer context of completing a study gives them new depth and the possibility to be much more lasting because of their clear utility.

This report is the students' opportunity to examine all the data they have, and find interesting aspects or connections. For some groups, particularly in a type three study, the amount of data collected by the partner can be overwhelming, and the process of sorting through all the variables to identify the most salient and interesting few is quite time-consuming and may result in numerous graphs, most of little interest. Although the students are learning inference at the time this report is due, they need not perform any inference for this report; it is enough to examine their data graphically and describe what they see.

For type four projects, this report and the second are much closer than for the other types. What is expected here is more of a progress report on the statistics; typically some agency cannot or will not deliver the statistics the students anticipated, and so they must find another source, and/or measure a slightly different quantity. This adjustment typically ripples through the rest of their analytical plan, requiring more adjustments.

Inference and Synthesis

The final report departs from memo format, taking instead the form of a technical report. Students are required to write a 50-word abstract for a report that must contain an introduction; sections on study design, description of data, and analysis; and a conclusion. The three interim reports are preliminary versions of the introduction, study design, and data description sections, respectively. The heart of the report, the analysis section, must include either appropriate use of inference to quantify strength of conclusions drawn from data, or an explanation of why inference is not appropriate for this study. The latter requirement is the only instance where the assignment is driven by a pedagogical goal that is not paralleled by a need of the community partner. However, understanding inference is a central goal of the course. It would be artificial (and sometimes impossible) to eliminate project topics that do not allow for inference. Instead, the students are expected to consider opportunities to buttress their conclusions by inferential analysis, and apply the inferential techniques they know. Sometimes (though rarely) a project has no need for generalization to a larger population; in this case the students gain an instructive experience with inference by explaining why it does not apply [11]. More frequently, the data has been collected without randomization, and the limitations of inferential conclusions in this situation must be conveyed.

The synthesis process for drawing conclusions with meaningful implications for the partner and supporting them by inferential analysis is the most demanding requirement placed on the students. It combines use of the most sophisticated statistical ideas presented in the course with discernment of connections between data and a rich external context. Students have opportunities to discover that data have meanings that they do not anticipate and cannot identify until they have considered the situation of their partner. This synthesis appeals to students, especially many who do not find the abstraction of mathematics compelling, and so amplifies the value of the course for them.

The structure of the final report encourages students to develop their ability to write technical material. The abstract's unique constraints and requirements stretch students' ability to distill the most important themes from their work. The greatest drawback in this report is lack of opportunities for feedback and improvement of the final document. Although students meet the WA with a draft of the final report, they don't have the benefit of revising to incorporate comments from the instructor. The analysis and conclusion sections in particular could be improved if there were earlier drafts, but there isn't enough time for presentation of statistical ideas, their incorporation in the report, and revision.

Teamwork and Assessment

Doing quantitative work in the context of a group and with an external entity relying on the outcome of the team's effort is a complex setting socially as well as intellectually. The course is designed to offer students plenty of time to identify potential partners for their semester project. The dynamics behind team selection and delegation of responsibilities is complex and one of the most difficult aspects of managing the course.

During the first couple of weeks, there are several in-class collaboration opportunities. The instructor insists that students work in new groups for each new activity so that students get plenty of exposure to their classmates, who have a wide range of academic and mathematical backgrounds. Students are responsible for the group they choose to work with, and it is important to find teammates whose expectations match one's own so that the group will function smoothly on the project, which comprises a quarter of the grade. While some students heed this admonition, others do not and seem surprised when they discover that they have chosen to work with a student with drastically different expectations (either higher or lower) than their own. To compensate for resulting inequities in contributions to the project, each student is asked to provide a final entry in the project journal to be read by only the instructor. This entry can include any summary comments on the value of the project in learning to apply statistics, and is expected to include an assessment of the contribution of every member of the group to the project (including the writer's contribution). This serves as motivation for students with lower expectations to rise to demands of group mates with higher expectations, and has served to increase equity in the distribution of work within groups.

Project grades consist of a team portion (based on the reports themselves) and a smaller personal portion (based on journals, including the assessments by group mates). The instructor reviews the journals in assigning the personal portion, looking for evidence of important ideas in the journal itself as well as attempting to create a coherent idea of the peer assessments. This portion of the grade can be either positive, for students who have made appropriate contributions to the projects, or negative, for students who have been consistently delinquent. This grading practice has not eliminated the tendency to freeload, but it has made the behavior less prevalent, less extreme, and less rewarding.

The project has increased in size as the course has evolved [10], and the course, which originally included three hourly exams and a two-hour final, now has only two exams and a one-hour final. Nonetheless, there are adequate opportunities to assess students' textbook knowledge of all material covered, and the final report plays such an important role in the students' learning experience that it seemed redundant to have a two-hour final that artificially distinguishes between the classroom/textbook presentation of concepts and their application. Although the project seems to result in an overall increase in time commitment, most students seem willing to make that commitment.

The Writing Associate

Reports and journals are joined through meetings with a "Writing Associate" (WA), an undergraduate provided through the College Writing Program to assist students by clarifying their message and voice. The WA meets with teams to discuss a draft of each report a few days before it is due. Journal entries chronicle development of the student's understanding of the project, and are submitted to the WA 48 hours before the meeting concerning the draft. Keeping a journal offers regular, structured opportunities for student reflection. Reading journals provides the WA with context in which to read the draft report, identifying both issues and sources of important ideas.

The relatively recent addition of a WA has resulted in far-reaching improvements. Because the WA reviews reports before they are due, it is not surprising that the reports show more careful attention to assigned goals, a clearer sense of audience, better diction, and fewer errors of syntax or grammar. Because the final report includes a large segment that revises earlier reports, this material is revised three times rather than simply once, with commensurate improvement. However the secondary effects of the WA are equally impressive.

Before the involvement of the WA in the course, students were required to submit reports to the instructor for review. This was to ensure that the reports were adequate, and reports often required rewriting before they were submitted to partners. As a result, these reports typically did not reach partners until well after useful feedback was possible, leaving partners out of touch. Because the WA can identify reports that lack clarity and/or basic features of the assignment, memos are suitable for partners to read once they have been revised in response. Thus students submit copies of reports to the instructor and partner simultaneously, and as a result partners are more aware of the progress of projects and can offer feedback to refine or revise goals. Long-time partners have expressed appreciation for timely updates, and improved communication is apparent in more recent projects.

Long Term Relationships and Effects on Partners

As described above, projects grow out of relationships between the students, the college community outreach center, the instructor, and community partners. Each semester presents a different array of part-

Type 3 Project
Identifying difficulties for a homeless shelter

A homeless shelter believes that it is experiencing an increase in admissions of functionally disabled people (FD). Analyses of admissions data for the past several years revealed several trends, but no trend in the rate of admission of FD. The next semester, a new team looks at more data and finds that, although there is no trend, the admissions for FD are more variable than other categories. The students recommend that the homeless shelter coordinate with other nearby shelters to even the load of FD.

ners with varying degrees of familiarity with data collection and statistics. Occasionally data collected by community partners are of poor quality, the most common problem being incomplete data sets. This is typically the result of staff or volunteers who have many responsibilities and don't see value in recording the data requested. In fact, giving the organization the sense that the data collected are valuable and might be used for its own advancement is an important secondary benefit of these projects. When subsequent studies are performed, the quality of data often improves. For example, a local low-income day care facility requested analysis of its attendance records to aid in staffing decisions. The original data were riddled with omissions, ambiguously labeled, and cryptic. The group chose one set of records, and worked to transform it into useful data. In the end, they were able to draw useful conclusions about staffing needs. More importantly, the center now gets more consistent records, since there is understanding that these records have utility.

Sometimes data collected by partner organizations are limited in other ways. For example, a common benefit for partner organizations in type three and especially type two projects is the improvement of their survey instrument. When students make recommendations to change a questionnaire and then can demonstrate the advantages of the clearer questions or improved response structure they have suggested, it offers special value as

Type 4 Project
School's tenacious drop-out rate explained

A local volunteer coordinator/community mobilizer wanted an assessment of the local public school's dropout rate. The students obtained summary statistics from the local school district, the state education department, and the federal census bureau. They showed that for the local school district, the proportion of students not graduating from high school is higher than the state average and declining more slowly than desired. Further investigation demonstrated that the problem could be a Simpson's paradox. The state dropout rate for Hispanics was much higher than for other major ethnic groups, and the local school district is experiencing a more rapid influx of Hispanics than the state as a whole. Thus, while the school district is improving the retention rate for all ethnic groups, the rapid increase in the one group with the highest dropout rate masks their progress. The community mobilizer plans to use the report to encourage local community-based organizations to tailor programs to address the educational needs of the growing local Hispanic population.

well as a potent learning opportunity for all. A study of participant satisfaction with the local YMCA aquatic program allowed one enterprising group to compare responses on the form previously used by the Y with a survey designed by the students. They were able not only to summarize the generally positive responses of swimmers, but also to demonstrate the advantages of the questions and response structure they proposed.

Difficulties Applying Statistics

When the report structure described here was instituted, the hardest report to motivate was the third. The students seemed to think that once the data was displayed and described, their work would be done; this would be their final report. In fact, identification of important aspects of the data and interpreting their meaning for the community partner is the capstone of the project. Until students have transformed data into workable conclusions about the world, they have not applied statistics. The distinction between observing features of a distribution and drawing conclusions based on these features is now a theme in the course, and students have little question about the role of the third report in development of their work.

Student Reaction

Not all students are prepared for the experience that this class offers. Some students cling to a preconception that this course cannot possibly offer value commensurate to the work demanded by semester projects. There is a strong (but not perfect) association between this opinion and the strength of students' mathematical foundation. Skeptical students often form groups with stronger, more motivated students; some learn statistics in spite of themselves. Others fail to pull their own weight. Once every couple of semesters or so a group comprised entirely of students unwilling or unable to do the work fails the semester project. These are typically students who are failing or nearly failing the course, and so it is natural that, having not grasped the basic ideas being presented, they are unable to apply them.

The quality of projects varies over the gamut from outstanding to unacceptable. A poorly done project has never ended the relationship between a partner and the course, but it can strain the relationship. In our experience community partners are understanding when some students show a weak commitment to a course, and they will work around a poor report. Often a project in a subsequent semester can address an outstanding issue; this most often arises from a conscientious group identifying need for further study.

The most common comments made by former students are that they are impressed with the utility of statistics or that the project taught them a lot about the issue they studied. Both of these seem to indicate that the projects play a critical role in cementing the students' understanding of statistics and in offering them motivation to encounter the material more deeply and learn it more lastingly.

Conclusion

The opportunity to learn statistical ideas in a more-or-less traditional classroom and textbook context and then immediately apply them to a statistical study of a topic of the student's choosing has powerful advantages. The students see mathematical ideas in multiple contexts and so find that the applicability of mathematical ideas is much broader than many had hitherto believed. The immediate applicability of their ideas becomes more apparent to the students as the semester progresses until, when the class arrives at the capstone concept of inference, motivation derived from applicability is felt by all but the most detached students. The civic dimension of the learning experience has great value but comes as a natural complement of the statistical learning experience. The result is a course that positions statistics squarely in the midst of a liberal education, with its utility and pervasiveness apparent, and its connections to the life of an educated citizen clear.

The authors gratefully acknowledge the efforts and cooperation of all the community partners who have participated in our course over the past five years.

References

1. Jon E. Anderson and Engin A. Sungur, "Community service statistics projects," *The American Statistician*, 53 (1999) 132–6.
2. Carmen Batanero, Juan Godino, & Antonio Estepa, "Building the meaning of statistical association through data analysis activities," *Proceedings of the 22nd conference of the International Group for the Psychology of Mathematics and Education*, University of Stellenbosch, Stellenbosch, South Africa, 1998.
3. Johnny L. Duke, "Service-learning: Taking mathematics into the real world," *The Mathematics Teacher*, 92 (1999) 794–796, 799.
4. Joan Garfield, "How students learn statistics," *International Statistical Review*, 63 (1995) 25–34.
5. Brian Jersky, "Statistical consulting with undergraduates: A community outreach approach," *Proceedings of the 6th International Conference on Teaching Statistics*, International Association for Statistical Education, Voorburg, The Netherlands, 2002.
6. David A. Kolb, *Experiential Learning: experience as the source of learning and development*, Prentice-Hall, Englewood Cliffs, New Jersey, 1984.

7. Clifford Konold, "Issues in assessing conceptual understanding in probability and statistics," *Journal of Statistics Education*, 3 (1995) 1.
8. National Research Council, *How People Learn: Brain, Mind, Experience, and School*, Expanded Edition, John D. Bransford, Ann L. Brown, Rodney R. Cocking, eds. National Academy Press, Washington, DC, 2000.
9. Marilyn Nickson, *Teaching and Learning Mathematics: A Teacher's Guide to Recent Research*, Cassell, London, 2000.
10. Rob Root and Trisha Thorme, "Community-based projects in applied statistics: Using service-learning to enhance student understanding," *The American Statistician*, 55 (2001) 330–335.
11. Alan Rossman and Beth Chance, "Teaching the Reasoning of Statistical Inference," *College Mathematics Journal*, 30 (1999) 297–305.
12. Jon Singer, Ronald W. Marx, and Joseph Krajcik, "Constructing extended inquiry projects: Curriculum materials for science education reform," *Educational Psychologist*, 35 (2000) 165–178.
13. Trisha Thorme and Rob Root, "Community-based learning: Motivating encounters with real-world statistics," *Proceedings of the 6th International Conference on Teaching Statistics*, International Association for Statistical Education, Voorburg, The Netherlands, 2002.

About the Authors

Char Gray is Director of the Landis Community Outreach Center at Lafayette College and leads community engagement work on campus. She also is responsible for working with faculty to integrate service into the curriculum. Her research focuses on development of problem solving skills among students engaged in service.

Rob Root is an Associate Professor of Mathematics at Lafayette College. Root investigates community-based research to teach elementary statistics more effectively. Other scholarly work focuses on biomechanical models of fish.

Trisha Thorme is Assistant Director of the Community-Based Learning Initiative at Princeton University. She facilitates the collaboration of students, faculty, and community partners on community-driven research projects and works with faculty to integrate such projects into courses throughout the curriculum. An anthropologist, Thorme's recent work explores how students, faculty and communities change as a result of community-based research.

3.3

Integration of Service-Learning into Statistics Education

Engin A. Sungur, Jon E. Anderson, Benjamin S. Winchester
University of Minnesota, Morris

Abstract. We present ways in which we have improved our statistics curriculum by making connections with our community through service-learning projects. We focus on three primary approaches: course structure, course content, and bringing in outside contacts and experiences. Course structure aspects include course projects and assignments and the creation and integration of our Civic Engagement Workbook. Under course content we present examples to illustrate learning objectives through a community connection. Collaboration involves units both within and outside the university, the incorporation of consulting, and student feedback and reactions.

Introduction

The approaches described in this essay represent one aspect of our efforts to improve our introductory statistics courses within a broader spectrum of changes and innovations along lines described in [1–4]. We view these approaches as complementary to our existing instructional framework. Some of the preliminary work in this particular area can be seen in [5–7].

It is our belief that students are more interested in the class, and thus learn better, when course content and course structure involve a variety of interesting problems, situations, and contexts. Incorporating notions of community into the course is one way to accomplish this goal. In addition to our own teaching experiences and those of other instructors, there is considerable theoretical support for our assertion. Previous work in cognitive theory [8, 9] suggests that exposing students to problems and concepts in a variety of contexts facilitates learning. Encountering statistics through notions of community also enables students to bring in their own community experiences and interests to support the constructivist view of learning new material [10].

Background

We are located at the University of Minnesota, Morris (UMM), a public liberal arts college and a part of the land-grant University of Minnesota system. Our campus is located in Morris Minnesota, a rural community of about 5000 residents located in west-central Minnesota. We have approximately 1900 undergraduate students with academic programs in the sciences (including major and minor programs in statistics), social sciences, humanities, and education. Our students typically come from smaller or mid-sized communities in the Midwest, or from the Minneapolis– St. Paul metropolitan area. We also have a substantial number of international students, and students from many other states.

The statistics program at UMM has shown substantial development in the recent years. First, it was a part of the Mathematics discipline and offered only an area of concentration in statistics. More recently it became a separate discipline, which is unusual for a liberal arts college of this size. Our curriculum is outlined in Figure 3.3.1.

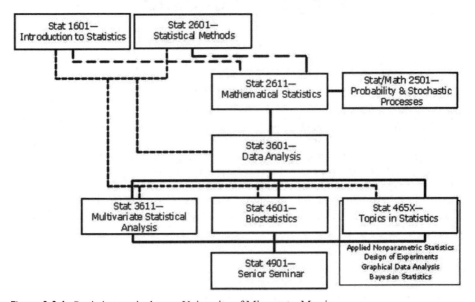

Figure 3.3.1. Statistics curriculum at University of Minnesota, Morris.

Our introduction to statistics course is designed as a service course for a variety of academic majors, and a way to fulfill a quantitative general education requirement. We currently use the text by Moore and McCabe [11]. The course is taught by three faculty members in the statistics discipline, and we typically enroll about 360 students in a two semester academic year.

Course Structure

Although the integration of service-learning in statistics courses differs by instructor, we have found the general approach given in Table 3.3.1 very useful. A community awareness component can be integrated into the course structure in many ways. Classroom examples, homework and exam questions, learning checks, required or optional chapter

Table 3.3.1. Stages of Service-Learning Implementation.

STAGES	ACTIVITY	NOTES
PRELIMINARY STAGE	1. Meet with city officials and community organizations. 2. Attend focus groups to address issues and define problems. 3. Data Collection and organization. 4. Present data to Professor.	This phase can be done primarily by the student coordinator or through a campus organization such as Center for Small Towns.
COURSE INTEGRATION	1. Determine classroom problems related to preliminary findings. 2. Assignment of group analysis data. 3. Prepare handouts and data disks.	This phase can be completed by the project coordinator and faculty.
STUDENT ANALYSIS	1. Administer pre-service-learning survey. 2. Examine primary values of data, and its relation to course objectives. 3. Random sampling in their area of analysis in the community. 4. Enhancement of model- building techniques, followed by creation of housing and transportation models. 5. Optimization of model, incorporating community needs and priorities.	This phase consists of work of students, faculty, the project coordinator and community organization representatives.
REPORT GENERATION	1. Students, in conjunction with the faculty, report on their findings concerning community problems, data summarization and analysis, model building, and interpretation.	
POST SERVICE-LEARNING REFLECTION PERIOD	1. During this period, the standard post-test evaluation forms from Campus Compact are used for assessment. 2. Publication of the academic and communal benefits of service-learning.	

or course projects, and classroom discussions are a few. It could be a required or optional part of the course. Also, such a component may be introduced to the course directly or indirectly.

The distinction between the direct and indirect community awareness component is driven by how the problem is identified and presented. In some cases a community organization directly introduces and brings the problem; in other cases a third party, typically a faculty member from another discipline, encounters a community-related problem. For example, a faculty member from the biology discipline studied how to best manage the remaining tracts of prairie to maintain diversity, prevent invasion of exotic species and woody vegetation, and promote their use by native wildlife. She was also interested in whether native prairie can be used to provide an economic benefit to its owners (e.g., through its use as pasture for livestock) without radically changing its species composition, or eroding its ability to support wildlife. The data collected by the investigator were analyzed fully by Applied Nonparametric Statistics students and partially by Introduction to Statistics students. This is an example of using an indirect community service component in a statistics course. Table 3.3.2 lists these components.

Table 3.3.2. Course components and service-learning.

Course Component	Service-learning Approach	Notes
Homework	Direct and indirect	Required
Exam	Direct and indirect	Required
Learning Checks	Direct and indirect	Optional
Quizzes	Indirect	Required for some instructors; optional for others
Examples	Direct and indirect	Not applicable
Chapter Projects	Indirect	Optional
Course Projects	Direct and indirect	Required for some instructors. Optional for others (5% extra points offered)
Classroom Discussions	Direct and indirect	Not applicable

Classroom discussions can involve the presentation of a problem by a community member and/or principal investigator. These could be a city official, a faculty member, a local business, or a community development representative. The information/data that is obtained can be easily integrated within the course structure. Even though students might be informed at the beginning of the class and on the course syllabus about the inclusion of a community awareness component, we prefer not to. Replacing what one does in our classes with examples and questions related to the community does not require a change in the policies and procedures of the course.

From a course procedure point of view, a community awareness component could be introduced at various levels. We could provide students everything they need, or on the other extreme they might formulate research questions and collect all the information needed. Thus, teaching approaches can range from practice-based learning to inquiry-based learning. These approaches and community awareness connections are summarized in Figure 3.3.2 and interpreted in our context in Table 3.3.3.

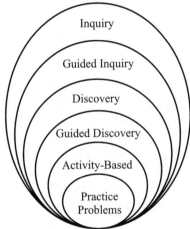

Figure 3.3.2. Diagram of teaching approaches.

Table 3.3.3. Guide to teaching approaches and community awareness components.

Teaching Approach	Community Awareness Component
Practice Problems/Drilling	Students are given a community related data set and asked to use predetermined statistical methods and techniques.
Activity-Based	Students are given a community related data set and asked to carry out activities related to the application of appropriate statistical methods and techniques to the data.
Guided Discovery	Students are given a community related problem and data and are asked to follow step-by-step instructions to select an appropriate statistical method or technique.
Discovery	Students are given a community related problem and data and are asked to select an appropriate statistical method or technique.
Guided Inquiry	Students are informed about a community related issue in general, and asked to follow step-by-step instructions on identification of a problem, on collecting or generating all required information/data, and analyzing the data by using appropriate statistical methods and techniques.
Inquiry	Students are asked to identify a community related problem, collect or generate all the required information/data and analyze using appropriate statistical methods and techniques.

Civic Engagement Workbook

In recent years we added another tool to our program to enrich our students' service-learning experience. This tool is the Civic Engagement Workbook for Statistics (see www.mrs.umn.edu/~services/cst/statbook/). We strongly believe the workbook will help us move to the higher levels of learning identified in Figure 3.3.2 and Table 3.3.3.

The workbook is maintained collaboratively by the UMM Center for Small Towns staff and statistics discipline faculty. The mission of the Center for Small Towns is to focus university resources toward assisting Minnesota's small towns by creating applied learning opportunities for faculty and students. Two courses, Statistics 1601: Introduction to Statistics and Statistics 2601: Statistical Methods, now have an integrated civic engagement component that involves the survey, analysis, research and reporting of rural statistical indicators. As members of these introductory courses, students are involved in the compilation of variables, the analysis of rural indicators, and the creation of a data book for people working in communities and counties across West Central Minnesota. The Center for Small Towns serves as the outreach mechanism for this project.

This workbook is used as a supplement to the course text. The primary data source for this workbook is the 2000 US Census. Both the Census Bureau and the Minnesota Planning data repository web sites, which contain a wealth of data about the area, are used throughout these problem sets.

Homework turned in through the use of this book is completed in an approved word processing software program. While the data may need to be written out on paper, the final results of the problem sets require a copy and paste from a statistical software package, such as SAS or STATCRUNCH, into a format that can be brought together with the students' results. The site we created includes maps used for sampling design purposes, links to the data sources, and instructions for the activities. The student reports are made available to the community organizations to be used for various purposes.

It is important to note that students are asked to locate and analyze data from the area surrounding the university in approximately a 100-mile radius. This approach helps them understand their communities better, increases their motivation, and makes the "interpretation of results stage" of a statistical analysis more meaningful.

Course Content

Another way to incorporate community data or community issues into statistics courses is through interesting examples, homework problems, projects, or exam questions. Several examples are provided in this section in order to illustrate our general approach and level of treatment.

During 1997, adult women seeking care were surveyed in eight medical clinics and 17 WIC program sites in nine counties of west-central Minnesota, Kershner [12]. In the survey women were asked about their experiences with abuse within the past 12 months (current abuse) and experiences before they were age 18. In class, a bar graph displays the percentage that responded having experienced the type of abuse in each time period. This graph, given below in Figure 3.3.3, reveals interesting aspects about violence against women. Students quickly observe that emotional/verbal abuse is the most prevalent form in our study, both before age 18 and in the past 12 months. Students also observe the large difference in the heights of the bars in the sexual abuse section of the graph. This typically generates discussion about the shockingly high extent of sexual abuse before age 18. The great disparity between the percentages experiencing sexual abuse currently and before age 18 provides an opportunity to discuss the idea of duration of exposure to risk in this study and other situations like the number of traffic fatalities over long holiday weekends versus a conventional two-day weekend travel period.

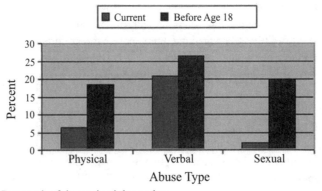

Figure 3.3.3. Bar graph of domestic violence data.

A discussion of confounding, or lurking, variables is another way to use these domestic violence data. The relationship between annual income and any type of current abuse shows that as annual income decreases the percent abused increases. However, after a careful adjustment for important variables like age, the relationship between income and abuse is no longer apparent. These data provide another example to motivate the concern we should have for potentially confounding variables. The example also provides an effective place to promote advanced courses in statistics where these kinds of adjustment techniques are taught.

This next example involves voting patterns. Morris is located in Stevens County, part of the rural, heavily agricultural part of west-central Minnesota. The county has a few small towns of under 800 residents and the population center of Morris with about 5400 people, including the University of Minnesota, Morris. The year 2000 presidential election results in Stevens County showed the same interesting pattern found in the rest of the United States: rural America tended to vote for Bush, urban America for Gore. Morris voting locations, the more urban locations in the county, tended to vote for Gore. The more rural township locations were overwhelmingly voting for Bush. After the election, these data were discussed in class, and then the class went to a national election website to observe the same rural versus urban voting behavior across the rest of the USA.

The 2000 US census data are becoming available for use at a very slow rate, especially for rural communities. In addition, rural communities do not have readily available resources to carry out the kinds of analyses that can lead to useful conclusions. Some of the students in our introductory statistics course obtained the 2000 census data for west-central Minnesota and analyzed it. Even though the project was optional (leading to extra credit), participation in this summer offering of the course was over 70%. Again, students expressed satisfaction in helping their community with a project that their community would not be able to do it by itself.

Some of the other service-learning projects that our students have worked on are: (i) climate and weather patterns in West Central Minnesota and the City of Morris, (ii) Pomme de Terre River characteristics related to river velocity, water levels, and soil erosion, (iii) climate impact on the Pomme de Terre River, (iv) the impact of Muddy Creek (the largest tributary) on the Pomme de Terre River, (v) non-game migratory bird inventory. This includes 10 classifications of bird categories in 12 wetland sites in 6 West Central Minnesota counties.

Our students in various courses have completed projects for the following clients: University of Minnesota's West Central Research and Outreach Center, U.S. Fish & Wildlife Service Morris Wetland Management District Headquarters, Stevens County Soil and Water Conservation District, UMM and Morris Public libraries, and the UMM admissions office. Some of our outside contacts have been arranged through the UMM Center for Small Towns, as mentioned earlier.

Course Survey Results

Toward the end of fall semester 2001, three sections of our introductory statistics course participated in a survey designed to examine many of the issues discussed in this paper. The three sections each had a different instructor, and there were 104 responses to the survey. The survey contained demographic questions concerning sex, academic year, and population size of home area. The remainder of the survey asked about the perceived importance of learning objectives and the interest students have in course content related to the concept of community.

High importance ratings were generally given to being able to apply statistical methods to coursework in their major subjects, which suggests that students were successfully recognizing the applicability of statistics in many different contexts and academic areas. Students also noted the importance of using statistics to address societal problems and everyday life experiences, and they indicated a strong sense of connection with their "home" community.

Being able to apply statistical methods to problems in sports, the campus community, the Morris community, and their home state were not rated so important by the students, but when we consider confounding variables, useful information may still be present. First, we note that the importance rating for sports is being influenced by a large number of female responses. Male students placed significantly more importance on applications to sports than female students. This has important implications for instructors, who should note that too much or too little sports content will probably not be as successful as some moderate amount.

We also found that the academic year of the student was significantly associated with the importance of using statistics to solve campus and Morris area problems. As might be expected, we found that older students who had been on campus longer were significantly more likely to respond that solving campus and local community problems is important. This finding poses interesting problems for an instructor in this course because students tend to be first and second year students who may not have yet developed a strong connection to the campus or local community.

Sections of the survey asked students about the interest they would have in examples or course projects relating to concepts of community. We formulated questions that concerned community on several levels: United States, California, vague (no location specified), Minnesota, Greater Minnesota (outside Minneapolis-St. Paul metropolitan area), Rural Minnesota, West-Central Minnesota, their home community, Morris community, and campus community. We found that students are most interested in examples and projects that reflect their sense of community. Examples about the United States, unspecified locations, or California are significantly less interesting than those for Minnesota or their home state. Students from smaller communities found examples relating to rural areas or Greater Minnesota particularly interesting. The data show that examples involving the student's home community, campus community, and Morris community to be the most interesting examples and project topics.

Conclusions

Our course survey results suggest that students are utilitarian, and find value in the usefulness of statistics to their lives. A relevant, successful course is more likely to occur if instructors are aware of student interests, concerns, and backgrounds. We have found that a student survey at the beginning of the semester helps give interesting data for course examples, and a way to learn about the class composition more generally. Course materials that cover a broad range of societal and liberal arts contexts would be appropriate for our students. The selection of examples and course materials would naturally change across institutions depending on the students and the educational mission of the course and institution.

The course structures and course materials relating to community issues and concerns described in this essay complement and facilitate student learning. These approaches are consistent with the findings of our student survey and our own teaching experiences. Our belief is that students learn best from examples and projects that are interesting and relevant to them. Community data and projects are valuable because they naturally will have some interesting academic, social, or scientific basis. Such data and projects also bring in a sense of ownership or understanding based on the community identities and affiliations the students bring to the course. During the implementation of this process we noticed the importance of the written and oral statistical communication skills of our students. In the next stages we will emphasize these skills somewhat more systematically.

References

1. R. Schaeffer, M. Gnanadesikan, A. Watkins, and J. Witmer, *Activity-Based Statistics*, New York: Springer-Verlag, 1996.

2. A. Rossman, *Workshop Statistics*, New York: Springer-Verlag, 1996.

3. G. Cobb, "Teaching Statistics," in *Heeding the Call for Change: Suggestions for Curricular Action*, Ed. Lynn Steen, MAA Notes 22 (1992) 3–43.

4. D. Moore, G. Cobb, J. Garfield, and W. Meeker, "Statistics Education fin de Siecle," *The American Statistician*, 49 (1995) 250–260.

5. J. Anderson and E. Sungur, "Community Service Statistics Projects," *The American Statistician*, 53 (1999) 132–136.

6. J. Anderson and E. Sungur, "Enriching Introductory Statistics Courses Through Community Awareness," *Proceedings of the Sixth International Conference on Teaching Statistics, Conference Papers, Developing a Statistically Literate Society*, Editor Brian Philips, ISBN: 085590-782-7, Electronic Publication, 2002.

7. R. Root, and T. Thorme, "Community-Based Projects in Applied Statistics: Using Service-Learning to Enhance Student Understanding," *The American Statistician*, 55 (2001) 326–331.

8. M. Lovett, and J. Greenhouse, "Applying Cognitive Theory to Statistics Instruction," *The American Statistician*, 54 (2000) 196–206.

9. J. Garfield, "How Students Learn Statistics," *International Statistical Review*, 63 (1995) 25–34.

10. E. Von Glasersfeld, "Learning as a constructive activity," in *Problems of representation in the teaching and learning of mathematics*, Ed. C. Janvier. Hillsdale, NJ: Lawrence Erlbaum Associates, 3–17, 1987.

11. D. Moore, and G. McCabe, *Introduction to the Practice of Statistics,* New York: Freeman, 2002.

12. M. Kershner, D. Long, and J. Anderson, "Abuse Against Women in Rural Minnesota," *Public Health Nursing*, 15 (1998) 422–431.

About the Authors

Engin A. Sungur is a Morse-alumni distinguished teaching professor of statistics at the University of Minnesota, Morris. He received his degrees from Carnegie-Mellon University (PhD and MS in statistics) and Middle East Technical University (MS in applied statistics and BA in city and regional planning). He is a national Learn and Serve scholar.

Jon E. Anderson is a Morse-alumni distinguished teaching associate professor of statistics at the University of Minnesota, Morris. Jon received his B.S. in economics (1983), M.S. in statistics (1987) and Ph.D. in biostatistics (1991) degrees from University of Minnesota.

Benjamin S. Winchester is the Coordinator of Data Analysis and Research at the Center for Small Towns located at the University of Minnesota, Morris. Ben received B.A.'s in Mathematics and an Area of Concentration in Statistics from the University of Minnesota, Morris (1995) and M.S. in Rural Sociology from the University of Missouri (2001).

3.4

Community Service Projects in a First Statistics Course

Debra L. Hydorn
University of Mary Washington

Abstract. This essay discusses three categories of community service projects that have been pursued in conjunction with an introductory statistics course. Topics related to successful project implementation are also discussed, such as reflection assignments, project management and promotion and tenure issues.

Introduction

I was first introduced to service-learning at a workshop conducted by the Community Outreach and Resources (COAR) Center of Mary Washington in the fall of 1996. The invited speaker at the workshop was Cecil D. Bradfield, a sociologist from James Madison University and co-founder of that school's Center for Service Learning. Dr. Bradfield described the service-learning component in two of his courses, providing details about the projects his students completed, along with a description of the journals they kept to document their learning experiences. What intrigued me most about service-learning from his presentation was its role in helping students to make connections between their course work and its applications in the community, together with its effectiveness in getting students involved in their own learning.

Also at this workshop the director of COAR described some service-learning activities at Mary Washington and distributed a handout of service-learning projects adapted from *Project Ideas for Combining Service and Learning for All Age Groups* from the National Center for Service Learning. Under each of the disciplines listed were two or more possible projects, each one offering a useful way for students to learn through helping their community. Among the disciplines listed, however, the closest I could find to my own was economics. Service-learning sounded like such a great idea I wondered why no one in my department was using it.

Among the materials I collected at that service-learning workshop was a flier describing a grant opportunity from Virginia Campus Outreach Opportunity League (VA COOL, now a part of Campus Compact) called the Faculty Fellows Program. I applied for and received this grant to support my efforts in defining and implementing a service-learning component in my Introduction to Statistics course for the fall of 1998. As a member of the first group of Faculty Fellows, I met several times during the following year with the other Fellows and with the members of VA COOL to learn more about service-learning and about the components of a meaningful service-learning experience. The Faculty Fellows also met at the end of our fellowship to share our experiences with each other and to describe our plans for sustaining our service-learning projects.

Since then I have incorporated service-learning into at least one of my Introduction to Statistics classes each semester. I have worked with a number of different agencies, all of which have greatly appreciated the assistance my students have provided in organizing and presenting quantitative information about the agency and the individuals they serve. I have experimented with incorporating service as a required component of my classes or as an option for a required data analysis project. Faced with the task of finding enough service activities for each one of my students, I have chosen to offer one or two service projects as an option each semester, instead of making it required. In this chapter I will describe the types of projects my students have engaged in as well as the participating agencies. I will also discuss some implementation issues that have not been addressed at length in the literature on service-learning in statistics courses.

Types of Projects

The service learning projects that my statistics students have participated in over the last five years can be classified into three different types: client summaries, program assessments, and community surveys. Each of these project types provides different learning experiences for the students, depending on the agency involved and the student's choice of project tasks. The projects themselves typically require producing output similar to the output of various computer lab projects in my class. With the description of each project type that follows, I have included one or two examples along with a discussion of issues related to the ease of project implementation and the degree of student interest.

Client Summaries

Many of the projects my students have worked on were data analyses needed by the participating agency as part of a funding proposal or for a required report to a funding source. Community service agencies rely on grant support to be able to offer their programs, and a substantial part of demonstrating the required community need involves documenting whom the agency serves. These projects are designed to help the agency to understand better who their clients are and how well their needs are being met. Client information may need to be summarized by race and gender, by city or county of residence, or by the services each client has used. After having completed one or more computer lab projects, students can appreciate the amount of time and effort needed for someone at the agency to prepare this information. When introducing students to this sort of project, I remind them that the agency may not have the time or other resources to devote to preparing these reports, and that the agency may not even have a staff member with the necessary data analysis skills. Encouraging students to focus on the agency's needs helps to impress upon them the value of the statistical methods they have encountered in my course. Students can appreciate the agency's need for the statistics they provide to obtain funding for its programs or to understand better the community it serves. Two of these projects are described below.

Rappahannock Area Office on Youth

Among the programs run by this agency, the Youth Community Corps provides court-ordered community service activities for juveniles in the Fredericksburg, VA area. The program director was interested in learning if community service referral rates were increasing. He also wanted information about the number of hours of service completed, as well as how long it takes juveniles to begin and complete their service hours. He provided the raw data for 1998 through 2000. The variables that were analyzed are shown in Table 3.4.1, along with a summary of the statistics produced by the students who worked on this project.

Table 3.4.1. Summary of the Office on Youth Project

Variable	Statistics produced
Age of juvenile	Mean, median, range, standard deviation, histogram
Time to start service (days)	Mean, median, range, standard deviation, histogram
Number of hours ordered	Average by month and number of juveniles involved
Time to complete service (days)	Mean, median, range and standard deviation by number of hours ordered

Association for Retarded Citizens – Rappahannock

This agency provides a newsletter, seven programs, and advocacy to mentally retarded individuals in the community. The director was relatively new to the agency and indicated that she did not know the ages and locality of the individuals served by each of its programs. She felt that having this information would also help her to know how to serve people not currently being reached by the agency's programs. For each one of its clients, the director provided me with the individual's date of birth, sex, county of residence, and a list of the programs used. The students who worked on this project provided descriptive statistics and graphs for these variables for each of the programs as well as for all programs combined, as summarized in Table 3.4.2.

Program Assessments

Another type of project my students have completed involves compiling and summarizing the results of a program assessment survey conducted by a community service agency.

Table 3.4.2. Summary of the Association for Retarded Citizens Project

Data for each client	Programs
Age, county of residence, number of programs	• Best Buddies (provides activities) • Dental • Respite Care • Daybreak (provides activities) • Advocacy • Hart House (provides activities) • Equipment Connection (provides computers)

These surveys are typically administered to the participants of one or more of an agency's education programs and are an important component of program assessment. The purpose of these projects is to help the agency learn how well it is meeting the needs of the community and to determine if other services are needed. The agencies I have worked with had already developed and administered a survey to program participants, and usually needed data entry along with a summary of the responses. Some agencies, however, may need help in preparing a survey instrument itself. Whether or not a survey has already been created and administered, agencies usually have a good idea of what kind of information is needed from a survey, which makes it easier to determine the appropriate summary statistics and graphs that the students should produce. Like the client summary projects described above, these projects provide data analysis help for the agency and usually involve producing frequency distributions of the responses to each question. Sometimes the agency has surveys completed by participants both before and after an education program, so that comparisons between each set of responses may also be needed. Two example projects are described below.

Rappahannock Council Against Sexual Assault

I have had students work on several different projects for this agency, which provides education programs on sexual abuse and assault for middle school children as well as for law enforcement and criminal justice professionals. The agency also provides advocacy programs for victims of sexual assault. The first project my students completed for this agency was an assessment of the effectiveness of its Stop Abuse through Family Education (SAFE) program. After the education program, children at area middle schools were given a questionnaire designed to assess their knowledge of sexual abuse. The students in my class who worked on this project created and compared frequency distributions of the number of correct responses for each question along with determining the proportion of the middle school children who had four of the five questions correct. My students were also asked to make comparisons by school, gender, time of day, and month of SAFE presentation. Another project for this agency required the students to tabulate adult participants' assessment of the knowledge, preparedness and communication skills of a workshop presenter for the agency's Legal Outreach Program. This survey also asked the participants to indicate what additional presentations they would like offered on topics such as sexual harassment and sexual assault prevention. A portion of the survey instrument used by the agency is provided in Figure 3.4.1.

Please rate this program using the following scale:	1=Needs Improvement 2=Fair 3=Satisfactory 4=Very Good 5=Excellent
Knowledge of the presenter(s)	1 2 3 4 5
Preparedness of the presenter(s)	1 2 3 4 5
Communication skills of the presenter(s)	1 2 3 4 5
Usefulness of information presented	1 2 3 4 5
Quality of handouts provided	1 2 3 4 5
How would you rate this program overall?	1 2 3 4 5

Figure 3.4.1. Example Program Assessment Questions.

Fredericksburg Area HIV/AIDS Support Services

The outreach coordinator for this agency needed assessment results for the HIV and AIDS awareness programs she offers. She provided me with questionnaires for each program participant, completed both before and after one of her education programs. The questionnaire included a section of HIV/AIDS knowledge questions along with a section on indicators of behaviors associated with decreasing the risk of getting HIV. A portion of the questionnaire is given in Figure 3.4.2. She was particularly interested in determining the number of participants who indicated an increase in risk-reducing behaviors. The students who worked on this project created response frequency distributions and contingency tables for making before and after education comparisons.

1. You can get HIV by hugging someone who is positive.	True False
2. HIV only affects certain types of people.	True False
3. HIV is curable.	True False
4. Circle the four body fluids that can transmit HIV:	
Saliva Semen Tears Urine Breast Milk Sweat Blood Vaginal Secretions	
5. Circle which behaviors you currently do to decrease your risk of getting HIV:	
don't have sex use condoms all of the time have only 1 sex partner	
carry condoms reduce # of sex partners use condoms more often	
don't have sex while using drugs or alcohol talk with partners about protection	

Figure 3.4.2. HIV/AIDS Awareness Questions

Community Surveys

Most of the projects my students have completed have been of the two types already described. However, I have worked with one agency director who requested my students' help in administering a survey created to assess the community's knowledge and understanding of his agency's function within the community. This project is described below. For students who are interested in the workings of community agencies, this kind of project is ideal. From start to finish, they learn what is involved with data collection and analysis. They can then adapt what they have learned to help campus or other community agencies with similar projects. I have also had groups of students, inspired by the service projects being done by their classmates, request to do a survey of their fellow students for a campus organization to which one of them belonged. Since the college environment is a part of our students' community, I have encouraged this kind of service project each semester. One of these projects was done for the biology club, with the approval of the biology department, to determine the opinions of biology majors on future course offerings and undergraduate research projects. Another project was conducted for the College's honor council to assess students' opinions and knowledge of our honor system. I work with these groups of students to prepare their survey and to devise a sampling scheme. These projects require a lot more student involvement, since students must collect data in addition to preparing frequency distributions of the responses to each question. The purpose of this kind of project is to learn what the community knows about what the agency offers and/or what their opinions are on the organization's issues.

Friends of the Rappahannock

This agency provides activities associated with the Rappahannock River, including river and regional cleanups, birding and other wildlife hikes, and canoe trips and tours. According to its mission statement, the agency also "provides advocacy, restoration and education programs to promote the conservation, protection and enjoyment of the Rappahannock River." The Executive Director had drafted a brief recreational use and community perceptions survey that he wanted to conduct by telephone. The survey was designed to determine who uses the river and what causes of river pollution were considered a problem by the community. After reviewing the survey with the director to clarify the format and goal of each question, I cautioned him about the potential difficulties that could occur with having students do the data collection for him. I agreed to attempt to assign to this project only those students who understood and accepted the responsibility of collecting useful and correct responses for the agency. To collect the data, students conducted a systematic sample from a phone book to select the numbers to call and used guidelines I provided for properly carrying out a survey by phone. I collected the completed surveys, entered all of responses and then provided each student with all of the responses for a selection of questions to analyze. A portion of the survey is provided in Figure 3.4.3.

Issues for Successful Implementation

Recent articles describing successful service-learning projects in statistics courses have provided good examples of projects and important information for faculty interested in

1. During the past year, have you used the Rappahannock River for any type of recreation? (for example, canoeing, fishing, swimming, or just walking along its banks) ___ Yes ___ No

2. Regarding the health of the Rappahannock River as a whole: Is your impression that the health of the River is excellent, good, fair or poor?

3. What do you perceive to be the biggest future threat to the water quality and scenic values of the Rappahannock River?
   (a) Farming Operations
   (b) Commercial and Residential Development
   (b) Industrial Pollution
   (c) Something Else

Figure 3.4.3. Community Use and Perceptions Survey Questions.

incorporating service-learning in their courses. One article provides a useful model for incorporating service-learning and describes service projects in upper level courses [1]. Another offers a narrative chronicle of the authors' experiences in incorporating service-learning in an introductory course along with a useful time line for managing projects [3]. These and other articles have not discussed some of the critical issues, however, such as what sort of reflection assignments are effective for helping students identify and appreciate the learning that has taken place, although the same authors have addressed some of these themes in their contributions to the present volume. This question is also discussed below, along with issues of working with other faculty to promote service-learning, dealing with data, project management, fostering a commitment to service, and tenure and promotion.

Reflection Assignments

Research on service-learning has established the importance of reflection in helping students to create and maintain a connection between course content and its usefulness to the community [2, 4]. Reflection assignments are perhaps the most difficult aspect to do well for the kinds of projects I have described, as student involvement may take place over a relatively limited time span. Because a student's interaction with the project may require only a few hours dedicated to data analysis, combined with time spent in group meetings and visiting the agency to learn about the agency and the purpose of the project, keeping a learning journal is not a particularly useful reflection activity in these circumstances. Instead, I have used a variety of reflection assignments, with each one intended to help students focus on what they learned from their service experience. Because my course also includes a required writing component, the agency and project descriptions, along with the progress reports and output descriptions that I require, are a part of the reflection process. By carefully combining the types of tasks students complete, I make sure that they each have at least one writing assignment that is about the agency or purpose of the project, and one that is about the output they produced that summarizes the results and addresses the agency's needs. This combination of writing assignments promotes the important connections that students need to make between course content and the community service they are providing. At the end of the project, I also ask students to write a concluding statement about the value of their service activities to the agency. When I have a large group of students interested in working on a particular project, far too many to make the learning experience meaningful if they were to all work on it, I ask each student to write a set of personal goals he or she hopes to accomplish through the service-learning experience, as a means of applying to work on the project. Upon completion of the project, I then ask those students who were selected to work on the project to revisit their goals to assess how well they were met. Finally, in-class discussions on the progress of each project also provide a reflection opportunity, one that also involves the rest of the class and can also impact the learning experience of students who did not work on a service project.

Working with Faculty in other Disciplines

Shortly after becoming involved in service-learning, I became acquainted with faculty members in other disciplines on my campus who had a similar interest in incorporating service-learning in their classes. I was very fortunate to begin working with one colleague in the psychology department who wanted to incorporate more service-learning projects in her upper-level research methods courses. We wrote a proposal and received grant support from Mary Washington to work together to improve the service components of our courses and to contact agencies to find projects for our students. We also worked together with the agency directors to make sure the agency's project would be assigned to the right course. For my course, I am most interested in basic data analysis projects. My colleague, however, wants her students to gain experience in designing and conducting a research project to meet an agency's needs. With two of us to handle calls and visits to agencies, we can contact more agencies and have had more success in finding suitable projects for our courses than each of us working alone. Each fall we have written a letter addressed to the directors of community service agencies to introduce ourselves and describe the kinds of projects we hope to find for our classes. An example letter is provided in Figure 3.4.4. We recently revised the process of gathering information from agencies about potential projects by creating an on-line form for the director to access and fill out, as well as including a hard copy of the form with this letter as an alternate method for providing us with the details of their needs. Some of the example projects that were included in the letter were omitted because they correspond to projects that have already been described earlier in this section. Working with another faculty member has also provided an important resource for identifying potential agencies to contact. Each of us had thought of different kinds of agencies to contact that the other had not considered. A list of the different kinds of agencies we have contacted is given in Table 3.4.3.

Dealing with Data

Some agencies may not want to open their files to students to compile the data for the project. In these cases I visit the agency to collect the data, and I am sometimes asked to sign a confidentiality agreement. When possible I like my students to complete as many of the required tasks for a project as is possible. Data entry is one task I usually do

Dear Agency Director:

Over the past year, we wrote to area nonprofit agencies offering help with data analysis, survey design, data collection and/or report writing. We are working to incorporate Community Service Learning into our Statistics and Research Methodology courses. We would like to share with you some of the projects that our students completed this past year and to encourage you to contact us if you have a project that you could use our help on.

Service learning projects for our courses might consist of creating and administering a survey to evaluate the impact of your agency on targeted community members, an analysis of data you already have collected on the individuals served by your agency, and/or the creation of tables and figures to represent your data. First, we will work with you to design a project to fit your needs. Second, we will work with the students in our courses to efficiently and professionally carry out the project. In any project, the confidentiality of your records and client information will be maintained and students will have access to only those items you indicate. During the past year, our students worked on the following projects:

Rappahannock Big Brothers Big Sisters. Students tested the effectiveness of using various ways to recruit volunteers. They studied whether emotional material, factual material, or a combination elicited the most responses. After researching marketing techniques, they designed and posted recruitment flyers for the agency.

Thornburg Middle School Gifted Program. In collaboration with the director of the program, students tested how gifted students differ from students not identified as gifted. The director was interested in learning which areas of self esteem her students might need help with. UMW students researched and ordered an appropriate test of self esteem, sought approval from the principal and school district, obtained consent from parents, and administered the test to the students. They completed a statistical analysis and reported the results to the middle school.

If you have need for a data analysis or survey design for the efficient running of your agency or for the completion of required reports, we would very much like to speak with you. Please feel free to contact us at one of the numbers below.

Figure 3.4.4. Typical Letter to Agency Directors.

myself, however, to ensure that it is in a usable format and error-free. But when a survey is relatively short and the questions involve selecting one from a small set of possible options, having the students do the data entry is a reasonable option. I also ask one student to error-check another student's data entries, to make sure the data file is correct. I have often worked with agencies that already had their data collected in a database, which can be a big time saver as far as organizing the data for the project. However, the data file provided by an agency may not be directly compatible with the software my stu-

Table 3.4.3. Examples of Agencies Contacted.

General Type of Agency	List of Specific Types
Health Related	• Community Clinics • Counseling Services • Hospices • Nursing Homes
Youth Related	• Teen Centers • Career Clubs • Public Schools
Community Support	• Advocacy Groups • Shelters • YMCA • Fire and Rescue Squads • Charities • Religious Organizations
Special Interest Groups	• Support Groups for Women, Minorities, etc. • Environmental Agencies • SPCA

dents use, which may require time to further prepare the data before students can work on it. I have also had to spend a lot of time working with an agency to clarify their data entry errors. In these cases, it would have been much easier for me had the data been provided in a raw form.

Project Management

Although I provide my students with a detailed time line that provides deadlines by which I expect each project task to be completed, some project management problems can occur. When conducting a phone survey, some students can become very frustrated with the attitude and responses they encounter. One student exclaimed that "only nice people answer surveys." I can only imagine the kinds of responses she must have received. Despite providing students with a unique perspective, which should encourage them to question how well respondents of any survey adequately represent the population they are from, I think that this sort of experience is one that should not be a part of an introductory course. I have found surveys to be the most difficult type of project to manage as there are so many different tasks involved. With so much work involved in carrying out the data collection part of a project, students may feel that they are working a lot harder than their classmates who did not choose to work on the service project. Another problem that can occur with any of the types of projects I have described is when the agency's report deadline is early in the semester but the project is to be an assignment for the end of my course. In this situation, I encourage students who want to complete their project earlier in the semester to work on the service project.

Fostering a Commitment to Service

Providing statistical assistance to colleagues and members of the community is an important part of why I enjoy being a statistician. I am often asked for advice by faculty colleagues about the design and analysis of a research project that they or their students are pursuing. I also get calls from individuals outside the college community for statistical advice as well as occasionally to help them with the actual data analysis. Most of these projects do not require a lot of my time so the only condition I often make before providing help is that I be allowed to share their project with my students. I believe the examples of my statistical service activities that I regularly include in my lectures help to demonstrate to students the usefulness of statistics and also help to motivate students to choose the service option for their course project.

Promotion and Tenure Issues

Once you have made up your mind that service-learning is something that you want to do, you will have to establish service-learning as an important part of your teaching activities. Even in disciplines with a long established history of service-learning, colleagues may not value a faculty member's involvement in service-learning. From my earliest experiences with service-learning, I was surprised to hear that non-tenured faculty are often advised not to get involved with service-learning. As a faculty member at a small liberal arts college, I am very fortunate that my activities in teaching are valued by my colleagues. In this environment, I have found my colleagues to be respectful, if not supportive, of my service-learning activities. To help demonstrate the value of service-learning, I have my students fill out surveys at the beginning and end of the course that include questions about their attitudes towards statistics and service. Figure 3.4.5 includes a sample of the

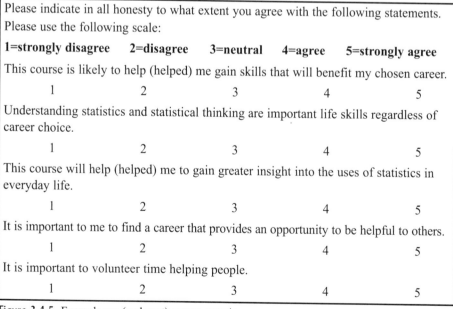

Figure 3.4.5. Example pre (and post) course questions.

questions I have used. I have given presentations on the outcomes of service-learning using the results of these surveys, and I also discussed the results in my tenure and promotion file. I anticipate that as more research on the outcomes of service-learning on student learning is conducted, the perceived value of service-learning will increase.

Additional comments

Providing service-learning opportunities for my students is definitely worth the time needed to find and prepare projects, and to guide students through the steps of creating and describing output for an agency. These projects do more towards showing students that statistics is an important tool for decision making and understanding variation than any other teaching method I use. They also allow me to share with my students one of the aspects I like best about being a statistician, being able to help others with the data analyses they need.

References

1. Anderson, J.E., and Sungur, E.A., "Community Service Statistics Projects," *The American Statistician*, 53 (1999), 132–136.
2. Mabry, J. Beth, "Pedagogical Variations in Service-Learning and Student Outcomes: How Time, Contact, and Reflection Matter," *Michigan Journal of Community Service Learning*, 5 (1998), 32–47.
3. Root, R. and Thorme, T., "Community-Based Projects in Applied Statistics: Using Service-Learning to Enhance Student Understanding," *The American Statistician*, 55 (2001), 326–331.
4. Schaffer, M.A. & Peterson, S., "Service learning as a strategy for teaching undergraduate research," *Journal of Experiential Education*, 21 (1998), 154–161.

About the Author

Debra L. Hydorn received her PhD in statistics from the University of Michigan in 1993 and joined the faculty of Mary Washington in 1994, teaching courses in statistics, finite mathematics, mathematical modeling and numerical analysis. She is a member of the American Statistical Association, the Virginia Academy of Science, and Faculty for the Twenty-first Century of Project Kaleidoscope. She was promoted to associate professor in 2000 and became the chair of the mathematics department in 2002.

3.5

Service-Learning Projects in Data Interpretation

Marilyn Massey
Collin County Community College

Abstract. Until recently, the service-learning projects available to mathematics students at Collin County Community College were limited to tutoring in the public schools or on campus. The search for new opportunities led to the development of four data interpretation projects covering topics in college algebra, discrete mathematics, pre-calculus for business and economics, and statistics.

Introduction

Service-learning is a teaching method combining service to the community with academic curricula in such a way that community needs are met and classroom learning is enhanced. Each service-learning opportunity requires a service project that is based on a community need and targets a specific concept in a course.

For several years, students have had the opportunity to participate in a service-learning project in mathematics at Collin County Community College (CCCC). Although student interest appeared high at first, actual participation did not meet expectations. Initially, the only service-learning opportunity available to these students was tutoring. There were numerous offerings through the local public schools, and students could tutor on campus through our academic support office. To tutor on campus, students must have earned an A or B in the previous mathematics course. Many students did not meet the grade requirement to tutor on campus or were not confident enough in their mathematics skills to tutor in the local schools. Others could not devote 1-2 hours a week during public school hours for 10-15 weeks to meet our 15-hour minimum requirement.

Although numerous opportunities are available for mathematics students to *volunteer* their services, finding true service-learning opportunities that support specific curriculum objectives has been challenging. This narrative will address a statistics project as well as three data interpretation projects developed as a result of a recent sabbatical.

Volunteer McKinney Center

The Collin County Community College Service-Learning Coordinator was contacted by the director of the Volunteer McKinney Center requesting assistance in the compilation of data obtained through a recent survey. Because this was the first time any survey had been undertaken by this organization, the results needed to be compiled and interpreted so as to present the information clearly and concisely to the center's board of directors. The coordinator knew this would be an excellent opportunity for statistics students.

The requirements of this project met several objectives of CCCC's statistics course. The course is a beginning course in non-calculus based statistics. It covers topics in sampling methods, histograms, frequency tables, measures of central tendency, measures of position, basic probability, probability distributions, z-scores, population estimates and sample sizes, hypothesis testing, and correlation and regression, among others. Although students would not be involved in statistical analysis, the project would give them a real-world view of questioning and collection techniques and the difficulties encountered when trying to interpret the resulting data.

Because of the necessity for a quick turn-around, every student in the class (22) was asked to participate for one to two hours instead of three or four students completing the entire project. Each student was given a packet of 10-20 response forms and a blank response form on which to compile the data for their packet. If a response form contained written comments, these comments were to be included in the compilation. Upon completion, two student volunteers reviewed each packet for accuracy and then compiled and interpreted the final results. A PowerPoint presentation was then created and given to the director for the board report.

Although individual students' involvement was minimal, each was required to respond to several questions about the survey. They reviewed the survey instrument and made suggestions for improvement. They also reviewed the collection technique (phone calls to randomly selected numbers) and also made suggestions for improvement. The students were asked to identify the statistics topics addressed and explain how the project clarified (or confused) their understanding of those concepts.

Students identified several areas that could be improved in the instrument in order to facilitate consistent, valid responses. For example in one area of the survey respondents were given only four means of response (very satisfied, satisfied, dissatisfied, very dissatisfied) while in another they were given six such choices (very satisfied, somewhat satisfied, satisfied, dissatisfied, somewhat dissatisfied, very dissatisfied). All students recognized that this lack of consistency could affect the results. Open-ended questions in surveys are occasionally necessary to obtain vital information, but every student recognized the difficulty in the reporting of such data. In addition, the students taking on the task of presenting the data learned the difficulty of interpreting data in the form of yes/no, scaled, numerical (How many hours did you volunteer?), and open-ended responses that had been included in a single survey.

The reflection process in this project was present, but not extensive. The students knew they had contributed to the success of the Volunteer McKinney Center's survey and felt empowered when they were informed that many of their suggestions were to be incorporated in the following year's survey. Although it cannot be verified that this project improved students' grades, their overall involvement in the course after the project was much more apparent in their attendance and class participation.

After the initial experience with this project, revisions were made in several areas that require students to answer specific questions. These questions are designed to focus their attention on the mathematics involved at each stage. The new guidelines are:

> When you have completed the tallying process, answer the following questions in paragraph form using complete sentences, proper English grammar, and include any additional information you think is appropriate.

> Did you have any difficulty tallying the data? Justify your response.

> Do you think enough people were contacted in order to get a valid representation? Mathematically, justify your answer.

> In your opinion, did this survey give the Volunteer McKinney Center the information it needed? Justify your response.

> In your opinion, is this a good survey instrument? Justify your response.

> Are there any changes you would make in the survey itself or the manner in which the data was collected? Please be specific and justify your answer with concrete statistical analysis.

This service-learning project is available to statistics students each year during the spring semester. Figure 3.5.1 shows the description of the project as it is presented for student consideration.

The reflection process also was modified and became the framework for all future reflective journals:

> When you have presented the data to the Volunteer McKinney Center, submit a copy of the presentation along with a reflective paper on your experience. You may

---

Course: Math 1342 Statistics
Objectives addressed: Data compilation and interpretation
Organization: Volunteer McKinney center (972-542-0679)
Contact person(s):
Number of hours required (per student): 14 + 1 hour mandatory orientation
Number of students requested: 5
Nature of project: Gleaning and interpreting data from surveys
Additional information: Knowledge of spreadsheets and PowerPoint may be necessary
Evaluation by organization? Yes

The Volunteer McKinney Center is a clearinghouse of information for the non-profit agencies operating in McKinney (and most of Collin County). Individuals call to volunteer and the center attempts to match their interests to agencies needing volunteers. Early each spring, the Volunteer McKinney Center surveys the public, persons in their volunteer database, and the agencies served. This project involves the tallying, compilation, and presentation of the resulting data. The initial tabulation can be divided among several group members to be rotated and checked by all other group members, but the final compilation and presentation to the Volunteer McKinney Center, either face-to-face (including charts/graphs) or in the form of a PowerPoint presentation, should be developed as a group.

Figure 3.5.1. Volunteer McKinney Center Project Information Sheet, Spring Semester.

address the following questions as a guideline, but your response should be in paragraph form using complete sentences and proper English grammar.

What were your expectations of this project? Were they met? Justify your response.
How did this project enhance your knowledge of statistics?
How did your knowledge of statistics enhance this project?
What specific changes/implementations will the Volunteer McKinney Center make as a result of the information you have provided them?
Do you feel you have contributed to the success/growth of the Volunteer McKinney Center? Justify your response.
In what way has this project allowed you to give back to your community?
How have you benefited from this project?
What was your overall impression/evaluation of this project?

The Volunteer McKinney Center project demonstrated that there were service-learning projects appropriate and available for mathematics students; the projects simply needed to be identified. This fact prompted the author's application for a sabbatical to research, identify, and/or develop additional service-learning projects. These projects are now being offered to students as options in various mathematics classes under terms defined by the individual instructors (for example, as a substitute for a test).

Project Descriptions

During the sabbatical the directors of all non-profit organizations working with the Volunteer McKinney Center were contacted about the possibility of projects suitable for mathematics students. Two directors had projects in mind when the first phone call was made and others provided time to meet and brainstorm. With their assistance projects were identified that could fulfill agency needs and meet curriculum objectives within various mathematics courses.

Three of these prospective projects are described in the following sections. The projects cover topics in college algebra, pre-calculus for business and economics, and discrete mathematics, but the unifying theme is that they all involve the interpretation of real data. Each project has a specific evaluation process requiring a presentation to the agency, and all require reflective journals following the guidelines of the Volunteer McKinney Center project described above.

American Red Cross

Two projects were identified within the American Red Cross organization in McKinney, Texas. In each, there was a need to organize existing data so as to allow staff members to make more informed decisions. Although the two projects are similar, they target different courses.

Students in the college algebra course would take existing data on donation campaigns, organize it, fit a function to the data, and use this function to predict the amount of future donations. They would also examine such information with respect to the specific month and number of days a particular campaign ran, the income received per day, and its cor-

---

Course: Math 1314 College Algebra
Objectives addressed: Functions, regression, predicting future values
Organization: American Red Cross (972-542-5642)
Contact person(s):
Number of hours required (per student): 14 + 1 hour mandatory orientation
Number of students requested: 3–4, preferably as a group
Nature of project: Taking existing data and fitting a regression function in order
to predict future values
Additional information: Knowledge of spreadsheets and PowerPoint may be necessary
Evaluation by organization? Yes

The American Red Cross serves thousands of people in Collin County. They keep records on each major donation campaign as well as money and goods donated at other times. They would like to know if there are trends in the donations and the overall effect of campaigns. This project will involve taking their existing data on the donations, using calculators or spreadsheets to fit functions to the data, making predictions about future campaigns, and presenting this information to them. Although the initial work of fitting an equation to the data can be done on an individual basis (the data rotated among the members in order to insure accuracy), the presentation is to be developed and presented to the American Red Cross (McKinney) as a group.

Figure 3.5.2. The American Red Cross Donations Project Information Sheet, Fall/Spring Semester.

relation to television and/or radio advertising. The evaluation process would require the student to submit copies of the data with corresponding regression functions along with answers to questions such as:

Did you have any difficulty finding the appropriate data for their specific goals? Justify your response.

Do you think there was enough data to effectively predict future outcomes? Justify your response.

What predictions are you making for future donations based on your results?

How accurate do you feel your predictions are?

Figure 3.5.2 describes the project designed for the college algebra course as it is presented for student consideration.

Students in the pre-calculus course would examine existing data on the costs incurred and revenue received for certain on-site programs (CPR training, for example) to determine the cost effectiveness of such programs. It is possible that outsourcing these training sessions could be more economical, but without data to support such a change, the American Red Cross would continue to conduct them. The evaluation process would require the student to submit copies of the data with the corresponding regression functions along with the answers to questions such as;

Did you have any difficulty finding the appropriate data for the specific goals? Justify your response.

Do you think there was enough data to effectively make a decision? Justify your response.

Based on your results, what recommendation will you make to the American Red Cross about the program you evaluated?

Course: Math 1324 Pre-Calculus for Business and Economics
Objectives addressed: Comparing cost/revenue functions
Organization: American Red Cross (972-542-5642)
Contact person(s):
Number of hours required (per student): 14 + 1 hour mandatory orientation
Number of students requested: 2–3, preferably as a group
Nature of project: Taking existing data on program revenues and participant numbers to deter-mine overall cost effectiveness of various programs
Additional information: Knowledge of spreadsheets and PowerPoint may be necessary
Evaluation by organization? Yes

The American Red Cross serves thousands of people in Collin County. They maintain records of funds generated and number of clients served on each program they provide (whether in-house or on-site). It is felt that some of the on-site programs (an example on-site program is CPR training) may not be generating enough revenue to cover the costs involved. This project will involve tak-ing the data of the funds collected per program to determine a revenue function, number of clients served per program to determine a cost function, and then compare them. A presentation of the resulting recommendations is required. Although the initial work of finding revenue and cost functions can be done on an individual basis (the data rotated among the members in order to insure accuracy), the presentation is to be developed and presented to the American Red Cross (McKinney) as a group.

Figure 3.5.3. The American Red Cross Cost Effectiveness Project Information Sheet, Fall/Spring Semester.

Figure 3.5.3 shows the project designed for Pre-Calculus for Business and Economics as it is presented for student consideration.

Collin County Committee on Aging

The Collin County Committee on Aging is responsible for, among other programs, the county's Meals on Wheels program. For this project, students would attempt to identify a more efficient route structure, if possible, to deliver hot meals to homebound seniors in Collin County. The current structure requires drivers to leave the central distribution point in McKinney to disperse to various locations within the county. At that point, each initial driver meets additional drivers to transfer the meals. Some proceed directly to the homes of seniors and others meet additional drivers elsewhere. Although the supervisor holds a degree in electrical engineering, he welcomes the opportunity to have students examine the present route structure and make recommendations. The project, as it is presented to students for consideration, is described in Figure 3.5.4.

Difficulties Encountered

After the projects were identified, one of the first difficulties encountered was the evalua-tion of students by the agency. CCCC had no instrument in place, and it was requested that a grade be given on impression. Many of the directors and contact persons were uncom-fortable with the lack of specific guidelines. With the aid of one agency contact and feed-back provided by others, an evaluation instrument was developed. This instrument allows the contact person to rate students in such areas as dependability, attitude, interest and

Course: Math 2305 Discrete Math
Objectives addressed: Graph theory
Organization: Collin County Committee on Aging (972-562-6996)
Contact person(s):
Number of hours required (per student): 14 + 1 hour mandatory orientation
Number of students requested: 3-4 as a group
Nature of project: Attempt to determine the best route for Meals on Wheels drivers
Additional information:
Evaluation by organization? Yes

The Collin County Committee on Aging provides the Meals on Wheels service in Collin County. From their central location in McKinney, they deliver hot lunches to approximately 450 – 500 homebound seniors throughout an area of roughly 900 square miles. Drivers from McKinney meet additional drivers at pre-determined drop-sites who then follow existing routes to deliver the food to the clients. Because it is necessary that the food arrive at the client's home as hot as possible, the Committee on Aging is always attempting to shorten the time between the central distribution center and the client. Students will work with the driver coordinator and incorporate graph theory in an attempt to determine the best routes for the drivers involved. As a group, students will present the results of the project, including any obstacles encountered, to classmates in a 20-minute presentation. Included in the presentation should be transparencies, slides, and/or hard copies of the original route structure as well as the proposed route structure. A mathematical explanation why the existing structure could not be improved should be included if appropriate. A group presentation to the Committee also is required.

Figure 3.5.4. Collin County Committee on Aging Meals on Wheels Project Information Sheet, Fall/Spring Semester

enthusiasm, organization, personal appearance, and delivery. For those projects involving direct contact with clients, students are also rated on understanding, empathy, and respect.

Another difficulty encountered is the number of students that can be accommodated at any given time by an agency. Whereas virtually any student wishing to tutor in a local middle or high school can find a placement, the majority of the projects discussed here can accommodate only a small number of students. For example, the *maximum* number of students that can participate in any one of the four projects presented in this essay is five, except for the one initial case where the class compiled survey results as a group.

To date, not all of these projects have been successfully completed by students. There are a number of explanations for this. Some students have had difficulty reaching the designated contact person. For example, in some cases, the contact person was no longer with the agency and a new one not yet identified. Some students did not complete the required paperwork quickly enough to be able to complete the project within the semester. In addition, not all mathematics instructors offer students the opportunity to participate in a service-learning project and for those who do, it is optional. The participating instructors generally include the project as an additional test grade and some students have indicated that the impact of this additional test grade was not worth the 15-hour minimum time commitment.

Conclusion

It is hoped that the projects presented here may provide ideas and motivation for other instructors to develop service-learning projects in their own mathematics courses. With prior knowledge of possible challenging areas in implementation, perhaps some difficulties can be avoided. Many directors of the agencies we visited have given an assurance that they are now aware of what is involved in offering a service-learning project and will keep a mind and eye open to future possibilities for student opportunities. There are of course many such agencies throughout the country. In addition to the American Red Cross, the following national organizations have specific needs in various communities: Goodwill Industries, Habitat for Humanity, Kiwanis or Rotary Clubs, the Salvation Army, and SPCA. The following county/city agency types may also offer additional service-learning possibilities: counseling centers, food pantries, homeless shelters, legal services, neighborhood clean-up/rebuilding projects, and senior activity or wellness centers.

References

1. Beckman, Mary. "Learning in Action: courses that complement community service." (Initiation Rights: Giving First-Year Students What They Deserve)." *College Teaching* 45, (1997): 72–75. In INFOTRAC. Cited 05 January 2000.
2. Bringle, Robert G. and Julie A. Hatcher. "Implementing service learning in higher education." *Journal of Higher Education* 67, (1996): 221–239.
3. Carpenter, Barbara W. and Jacobs, Jacqulin S. "Service Learning: a new approach in higher education." *Education* 115, (1994): 97–99.

Web Resources

1. American Association of Community Colleges: Best Practices in Service Learning, www.aacc.nche.edu/Content/ContentGroups/Project_Briefs2/bestprac.pdf
2. American Association of Higher Education—Programs and Partnerships: Service-Learning Project, Dec. 20, 1999, www.aahe.org/service/srv-lrn.htm
3. Campus Compact National Organization for Service-Learning, www.compact.org/

About the Author

Marilyn Massey has been a mathematics professor at Collin County Community College since 1991. During her tenure she has served as Campus Service-Learning Liaison, was voted Developmental Education Outstanding Faculty Member, and recently received a Center for Scholarly and Civic Engagement Pioneer Award for her contributions to service-learning.

Chapter 4

Educated-Oriented Service-Learning Projects in Mathematics

4.1

Perspectives on Education-Oriented Mathematics Projects in a Service-Learning Framework

Victor J. Donnay
Bryn Mawr College

Introduction and Overview of Projects

Service-learning can be defined as an activity performed by students for the benefit of the community while at the same time enhancing the students' own learning. In this chapter, there are nine essays that describe mathematics-related service-learning projects in which the activity performed by the students involves providing some form of mathematics instruction to the community partner. I will first give a brief overview of the projects, then discuss general issues related to such mathematics service-learning projects, and finally summarize with some future directions in which this type of service-learning could profitably move.

The service-learning activities in this chapter can be divided into three broad categories: tutoring (Craig, Dwyer, Hamman, Ridlon, Zang), developing and then teaching a hands-on, activity based lesson for K–8 students (Bonari and Farrer, McDowell, Myers), and having non-education majors learn about mathematics and mathematics pedagogy through teaching an ongoing series of lessons to high school students (Morse and Sabloff).

Except for Ridlon's America Counts tutoring program, all the service-learning projects are attached to courses. While the majority of these courses are targeted at elementary or secondary pre-service teachers, there are also descriptions of courses aimed at mathematics students who are not involved in pre-service programs (Hamman and Zang). For example, Zang describes how he has included a service-learning component in his finite mathematics course. Ridlon describes the implementation of the America Counts tutoring program at her university, a national program in which college students receive training in math tutoring and then tutor in a K–8 setting. College students can be paid their Federal Work-Study money for participating in the program. Mathematics Education faculty at her institution strongly encourage their students to participate in the program. The various components of these projects are outlined in Table 4.1.1.

The tutoring projects involve a range of community partners who receive the tutoring. In the tutoring projects described by Dwyer, Hamman, and Ridlon, the community partners are primarily K–8 schools. In Zang's project, the partners include an alternative high school program as well as community organizations, such as the Salvation Army and a Boys and Girls Club, that run after-school programs.

The service-learning projects of Bonari and Farrer, McDowell, and Myers involve the development and implementation by pre-service teachers of an activity based math lesson for K-8 students. The students observe in a K–8 classroom, then develop a math activity that is age appropriate and linked to state standards. The pre-service teachers then teach the lesson to the K–8 students, either as part of a Math Carnival (Bonari and Farrer) or a Family Math Night (Myers) held at a local elementary school, or as part of a Math Trail that the K–8 students explore at the university (McDowell).

In Craig's course, pre-service teachers run a weekly supplemental help session for students who are taking college algebra at her university. Thus the community partner is her

Table 4.1.1. Summary of service-learning projects in this chapter.

Author	Service-Learning Site	Pre-service Course	Discipline Course	General Elective	Tutoring	Lesson Development & Implementation
Craig	U. Central Oklahoma	Secondary			X	X
Dwyer	U. Tennessee	Elementary			X	
Hamman	Anne Arundel C.C.	Elementary	Developmental math		X	
Ridlon	Towson U.				X	
Zang	U. New Hampshire		Finite Mathematics		X	
Bonari and Farrer	California Univ. of Pennsylvania	Elementary				X
McDowell	U of Southern Mississippi	Elementary			X	X
Myers	U of San Diego	Elementary			X	X
Morse and Sabloff	U of Pennsylvania			Community Math Teaching Project	X	X

own university. The pre-service teachers develop and implement problem-solving exercises that make use of graphing calculators for the algebra students.

In Morse and Sabloff's course, the university students teach a weekly geometry lab to students at a local high school. Here the service-learning component is fully integrated into the course. One of the two weekly class meetings is held at the university during which the undergraduates learn about issues of mathematics pedagogy and about the geometry content they will be teaching. The second class session is held at the high school. Each undergraduate works with two high schoolers and leads them through that week's geometry lesson. In the following class meeting, the undergraduates reflect on their teaching experience as well as prepare to teach the next lesson. As a culminating experience for this course, the undergraduates design and teach their own lesson.

In most of the courses, the service-learning component is a required part of the course. In Dwyer and Hamman's courses, all students have to do some type of project and a service-learning project is one of the options. Dwyer points out the value of exercising some control over which students are allowed to undertake the service-learning option, as the students are representatives of the university and need to carry out their duties in a responsible fashion. In addition to undertaking the service-learning activity, the undergraduates are often asked to keep a journal of their observation and then write a reflection paper in which they make connections between their classroom learning and their service-learning experiences. Student performance on the service-learning component is evaluated and contributes to their grade for the course. McDowell's essay includes a very instructive rubric outlining how the service-learning component is evaluated.

Many of the essays have references to educational literature that gives evidence of the value of service-learning (see particularly Dwyer, Zang, and Bonari and Farrer). In addition, Craig, Ridlon, and Zang have done evaluations of the effectiveness of various aspects of their projects.

Drawing on the essays in this chapter, as well as my own experiences with service-learning, which include serving on the service-learning oversight committee of my institution and designing and teaching a service-learning course, I will discuss some of the questions that an individual, department or institution might have about service-learning. What are the payoffs for implementing service-learning—for the student, for the faculty member, for the institution? How much extra time and effort will it take and what are ways to keep the extra work at a manageable level? What types of challenges arise in implementing service-learning? What resources are available to support service-learning efforts?

Learning Issues

Mathematics Education Courses

Let us first look at the opportunities service-learning provides for improved learning. For pre-service teachers, the benefits of these types of service-learning projects seem self-evident. They will give future teachers the opportunity to have a hands-on teaching experience early in their college careers. The projects described here all happen early in a student's college career, well before the student teaching experience starts. By seeing how the material they are learning in class plays out in their service placement, the students connect theory with practice, thereby making the material tangible. By actively engaging the material, the students will learn it better than if they simply read, listen, and discuss without actually doing. In the pre-service classroom, the student is learning about the theory and methods of math teaching. By tutoring or by developing and implementing a lesson, the pre-service teacher gets to experience these ideas first hand which will lead to greater retention of the material. The students can bring back experiences from their service-learning activities to the classroom, leading to a more lively discourse and higher level of engagement. By having the opportunity to engage in teaching activities early in their studies, pre-service teachers can become excited about their chosen career path.

The National Council of Teachers of Mathematics (NCTM) standards [1], parts of which are now being implemented in state and national standards, call for active learning approaches; research indicates that such approaches produce improved learning. When they become practicing teachers, the pre-service teachers will be expected to engage their own students in such active learning approaches. It is an old adage that we teach the way we were taught. Thus to prepare pre-service teachers for the demands and expectations of their future work, it behooves us to provide them with active learning experiences such as service-learning provides. These experiences can be tailored to provide pre-service teachers with both active learning experiences and also with the opportunity to reflect on these experiences within the wider context of the theory of active learning.

For elementary pre-service teachers, there is an additional educational benefit to these types of service-learning experiences, as pointed out by Hamman, Bonari and Farrer, McDowell, and Myers. Many of these students are not confident in their own mathematical abilities. They come to their Methods of Math Teaching course with the attitude that since they will only be teaching elementary level math, which they have already learned themselves, they do not need to learn more advanced mathematics. Thus they are initially resistant to the whole premise of a math education course whose goals include teaching them higher level mathematical reasoning and the conceptual framework and logic behind the mathematical manipulations that are taught at the elementary level. Indeed many of those pre-service teachers initially believe that successfully learning algorithmic manipulations is the main goal of elementary school math classes. By serving as a tutor, seeing the issues elementary students face in learning mathematics and trying to answer their questions, the pre-service teachers realize that teaching elementary level mathematics is a much more complicated task than simply training students in algorithmic methods. They begin to see that if they are to be successful in their math teaching, they will need to deepen their own understanding of content. As their old attitudes change, the pre-service teachers can then become receptive to the material being taught in the methods course.

A common feature in almost all the service-learning projects is the use of either a journal or reflection papers to give students the opportunity to reflect on their experiences and their own learning. Engaging in such reflection activities, termed metacognition in the learning theory literature, has been found to be a key ingredient in helping students learn to take control of their own learning [2].

Although the essays in this chapter describe service-learning projects primarily for elementary pre-service teachers, secondary pre-service teachers would also benefit from the improved understanding and practical experience with active learning activities that a service-learning project could provide. The NCTM standards call for active learning in the secondary mathematics curriculum too. One can easily envision adapting elementary pre-service service-learning activities to fit the needs of secondary pre-service teachers. Students could tutor in a middle school or high school. One could carry over such activities as the Math Carnival, Family Math Night or Math Trail events to a middle or high school setting by using more advanced math activities and giving an age appropriate slant to the activities. The Association for Women in Mathematics sponsors Sonia Kovalevsky High School Mathematics Days at which colleges and universities bring high school girls to campus to engage in a day of math related activities. Secondary pre-service math students could design the activities for such a day, and funding for such initiatives might be available from various outside organizations.

Disciplinary Mathematics Courses

For students who are not pre-service teachers and are taking a disciplinary mathematics course, there is more of a question as to why service-learning would improve their learning since there seems to be less direct connection between the service-learning component and the content of the course. To see how service-learning can contribute to improved students learning in this context, we examine the service-learning tutoring model as it applies to two types of undergraduate math students.

The first are weaker math students who might be taking a math course to fulfill some type of quantitative requirement at the university, as is described in the essays by Hamman and Zang. By tutoring students at a lower level, for example K–6, the college students will have an opportunity to revisit basic material that they might be expected to already know. However, weak students often do not have complete mastery of this material, and the mastery that they do have is algorithmic. These students often do not understand the underlying concepts and principles behind the algorithms. By tutoring at lower levels and trying to explain the material to others, such students will have a chance to engage the material at a deeper level and strengthen their own understanding. I have heard students returning from an elementary level tutoring experience report excitedly, "I learned something today that I never understood before!"

Such eureka moments can help change students´ self-image of themselves as poor mathematics learners. Also, by having the opportunity to be the expert, albeit at a low level, they realize that their mathematical abilities are of use to and valued by others. This positive experience can make them more confident and motivated learners. Many weak math students report that they had liked and been good at mathematics until they reached a certain grade level or had a bad experience with a particular teacher. A positive tutoring experience can help remind these students of their innate enjoyment of mathematics.

The issue of quantitative literacy for all college students is presently gaining much attention [4]. The basic rationale is that in our increasingly technological and quantitative society, citizens need improved levels of quantitative skills to be able to make the informed decisions on which a democracy depends. Furthermore, people need high levels of quantitative skills to hold down the demanding technologically oriented jobs on which the vibrancy of our economy depends. A common theme heard in discussions of why many students perform poorly in math is that they do not see the relevance of mathematics to their everyday world. To help these students improve their math abilities, it is first necessary to help them see the relevance of mathematics. One approach being suggested in designing quantitative literacy courses is to include compelling real world applications in the course such as the statistics of the Presidential election count in Florida or of the spread of AIDS. Zang's essay suggests using service-learning as a way to add relevance to the material in a quantitative literacy course. Applying statistics or mathematical modeling to issues in the community would point out the relevance and applicability of mathematics. Tutoring and mentoring experiences could also serve to give students a sense of the relevance and importance of mathematics, particularly if the material involved had an applied focus.

For stronger math students, a tutoring experience with high school or college students could give them a good review of material and allow them to deepen their understanding of previously learned material. As an example of service-learning for discipline students, when I teach real analysis, I have a "math culture" requirement. The students either attend a number of colloquium talks or they engage in an ongoing tutoring activity for students in a lower level math course such as calculus. In either case, the students write reflection papers in which they look for connections between their math culture activity and the material in our analysis course. The tutoring option gives the students the chance to review the computation aspects of material that we are looking at from a theoretical perspective in real analysis.

Service Aspects

Now let us look at the benefits of service. There is a strong movement towards civic engagement in higher education as colleges and universities are being asked to explain how their activities are of value to society. There is a growing sense that rather than being ivory towers of academic isolation, colleges and universities should be involved in their communities. To prepare students to be contributing members of society, their education should include engagement with the community. Many states now have community service requirements for high school graduation, although these service activities do not necessarily include a formal tie with academic work and hence are not service-learning. Thus students are coming to college and university with previous service experience and a desire to continue these activities. Service-learning is a way to bring about community engagement while also linking with the fundamental academic mission of higher education. Given all the positive aspects of civic engagement, higher education administrators are generally supportive of service-learning. Indeed, savvy public relations offices now tout their service-learning activities as a selling point to prospective students.

For many students, the opportunities that service-learning provides to make a difference prove to be powerful, and in some cases even transformational, experiences. Many of the essays quote student statements that illustrate the personal impact that the service component produced. As a faculty member, I have felt a much greater sense of making a difference in my students' lives when I teach my "Changing Pedagogies" service-learning course, which I will describe later in this essay, than when I teach my predominately disciplinary courses.

For the recipient of the service-learning activity, the essays mention numerous positive benefits including having a college student acting as a positive mathematical role model and how such personal contact can encourage young students to raise their academic goals and think about attending college themselves. If the provider and recipient are both from groups that are historically under-represented in the mathematical sciences, then the impact on the recipient is even more powerful.

As Ridlon's essay on the America Counts math tutoring program shows, a university or college can use its service-learning activities as an outreach device to raise community awareness of the institution. Ridlon's implementation of the America Counts tutoring program offers a model of how one centrally organized service activity can have broad benefits across an institution. Ridlon has been Project Director for a program that involves a wide array of community partners, has received extensive outside funding, and had 148 tutors participating during 2002–03. By having such a centrally organized program, math and math education faculty who would like their students to do service-learning can partner with the America Counts program and save themselves much of the administrative legwork. As part of the America Counts program, students receive eight hours of tutor training before they start working and then ongoing training throughout the year. This level of training is well beyond what one could expect an individual faculty member to provide. Students who are eligible to receive Federal Work-Study as part of their financial aid package can earn their work study money through tutoring. The United States Department of Education requires that 7% of an institution's Federal Work-Study budget be used by students carrying out service-oriented community jobs. There has been talk in Congress about raising this level above 7%. Ridlon describes how her institution's Federal Work-Study allocation was increased by the federal government because of the large proportion of work study funds going to community service activities due to the America Counts program.

America Counts follows in the footsteps of the America Reads literacy tutoring program that operates on similar principles. At many institutions, tutoring programs are organized by the Community Service Office. People who run community service programs tend to be well versed in issues of literacy tutoring and as a result there are a large number of America Reads chapters nation-wide. There are far fewer America Counts chapters partly because many community service organizers are uncomfortable with mathematics and hence are hesitant to start up such a program. Mathematics faculty can play a vital role here; either by supporting and encouraging the efforts of the Community Service Office to organize an America Counts chapter or by leading such organizational efforts from within the math department, as in Ridlon's case.

Further Implementation Issues

One of the mantras of service-learning work is to expect the unexpected. The America Counts and America Reads programs were initiatives of the Clinton administration. When the Bush administration took over, these programs were terminated. Information and organizational materials for the program are still available at the web site, www.ed.gov/americacounts, but new materials are no longer being developed and training workshops are no longer being sponsored by the United States Department of Education. The Federal Work-Study provisions still apply and institutions can still undertake the very worthwhile America Counts activities; but it may be more of a challenge to get such a program set up.

As a faculty member, one of the aspects I appreciate most about service-learning is the opportunity it provides to combat the isolation associated with being in the academy.

When I do my mathematical research, I am working alone. I teach my courses without a lot of interaction with other faculty. But it is impossible to do service-learning and not interact with other people since one needs to have a community partner, and setting up the partnership invariably involves interactions with lots of people, both on campus and in the community.

I was asked to serve on my institution's service-learning oversight committee and through this met other faculty and staff who were involved in service-learning. This was an enthusiastic group of highly dedicated people who cared deeply about creating a more meaningful educational experience for their students and who enjoyed working to change the academic culture of our institution to this end. I always left the meetings of our service-learning committee energized and excited by all the good work we were doing and re-committed to carrying on with these efforts (in spite of the extra time and effort the service-learning activities took). I would encourage institutions to create a service-learning committee, involving a wide range of campus (and perhaps even non-campus) partners, as a way of looking for synergies and keeping the service-learning projects moving forward. If one follows the traditional faculty model of working in isolation, it is easy to become overwhelmed by the difficulties of service-learning. Being part of a group that meets regularly can keep one's enthusiasm up and leads to an exciting sharing of ideas and strategies.

Once one starts service-learning, and begins to think in terms of making partnerships, one discovers that this new way of looking at education leads to lots of opportunities for interactions and synergies. For example, Zang and Ridlon made partnerships with faculty and/or graduate students in statistics who then designed evaluation instruments to measure the effectiveness of the service-learning project. Ridlon describes how the data evaluating the performance of America Counts tutees will be used for a semester-long applied math project in an advanced undergraduate course. Taking this type of loop a step farther, one can imagine the situation of the service-learning component for a statistics course being to evaluate the effectiveness of the service-learning component of another course.

As a balance to the many benefits of service-learning, there are also challenges and difficulties involved. The most obvious is the extra time and effort it takes faculty members to design and implement a service-learning component, including finding community partners to work with and making the arrangements for placements. Fortunately, many colleges and universities now have some type of service-learning or community service office that can help. Visiting this office is a natural first step in the process of preparing to add service-learning to one's teaching. The staff in these offices can help one think about possible projects and issues of implementation and most importantly help with identifying and contacting community partners. Finding and making contact with the partners is often a very intimidating part of the project. Even if one's institution does not have a service-learning office, there are likely to be faculty on campus who are engaged in some type of service-learning activity and can be useful resources. At many institutions, faculty do not know what each other are doing, so it may take some searching to find these service-learning faculty. If a department wishes to have a large scale service-learning program, it would be helpful to have a point person for the project who, as part of their regular departmental duties (i.e., not as extra voluntary work), oversees service-learning (Dwyer, Ridlon).

In deciding on what type of service-learning activity to undertake, I would caution to start small, be successful, then grow. The projects in this chapter illustrate a range of complexity. Adding a tutoring component to a course might be the easiest way to start (Dwyer, Hamman, Ridlon, and Zang). A more complicated project would be having the students design a lesson and then implement it as part of a special event (Craig, Bonari and Farrer, McDowell, and Meyers). Finally a yet more ambitious approach is to totally integrate the service-learning component into all aspects of a course (Morse and Sabloff).

Faculty often feel that their courses are jam-packed as is and do not see how they could fit another component (service-learning) into the course. They also do not want to require an unfair time commitment from their students. The essays address this issue, showing ways to integrate service-learning activity into the curriculum so that it teaches important topics of the course and thereby can be used to reduce the class time previously dedicated to such topics.

There are a myriad of implementation details to address; the essays give lots of practical advice in this regard. Over time, managing the administrative aspects of the service-learning project will get easier but since aspects of the community partners' situation often change, there are always new issues to deal with. The essays stress the importance of soliciting regular feedback from all participating groups so as to identify problems and modify the program accordingly.

Institutional Issues and Funding

With the growing interest in service-learning, there are funding opportunities available to support service-learning from foundations and various other funding groups. A good place to start looking for funding is with the organization Campus Compact (www.compact.org), a national coalition that is committed to the civic purposes of higher education. To expand and formalize our service-learning activities, my institution received a three-year grant from a foundation. An institution's development office might also be able to find individual donors who would be excited about the opportunity to support such a worthwhile program. Such funding can be used to give faculty summer salary or release time to develop service-learning activities or to hire support staff that will assist in the implementation of service-learning. Even if one's institution does not have targeted service-learning funding, a chair, dean or provost often has funds for innovative curriculum development. In looking for administrative support for service-learning, be sure to point out the public relations and recruitment benefits to the university of such activities. McDowell's essay shows how the Math Trail activity can be used to advertise the strengths of an institution to the next generation of prospective students. If the administration is itself supportive of service-learning, a mathematics department can earn political capital with the administration by participating in service-learning. Although it is possible for a determined and motivated faculty member to initiate service-learning without formal support from the institution (Zang), such support is vital if service-learning is to become institutionalized and outlive the efforts of the heroic individual.

How will the extra time faculty members spend on service-learning affect their chances for promotion and tenure? Are efforts put into service-learning activities valued as scholarly contributions to the educational enterprise by those making appointment decisions? After all, time spent on teaching is less time for research. Zang points out that there are opportunities for research imbedded in service-learning activities. For education faculty, I would expect that a department would strongly value service-learning activities since they make such a worthwhile and obvious contribution to the program. Although the reaction to service-learning in the mathematics discipline might be less supportive, there seems to have been a shift in the academic culture over the past decade with more emphasis being placed on the educational role of mathematicians and on the importance of pedagogy in mathematics content courses. The discussions around calculus reform have contributed to this change, as well as such programs as the NSF funded VIGRE initiative (NSF 01-104) and the Mathematical Association of America's Project NEXT.

Recently, I served on a mathematics search committee at another institution. As the position was at a liberal arts college at which teaching is highly valued, I expected to see the applicants stress their interest in teaching. But I was nevertheless struck by the high level of pedagogical engagement the candidates demonstrated. Included among their activities were teaching reform calculus, teaching using technology, having students do projects and make presentations, using group work, supervising undergraduate research, and participating in seminars on pedagogy. To stand out from this talented pool, a candidate would have to demonstrate a very special commitment to teaching. It turned out that the candidate who was eventually hired had undertaken extensive service-learning activities. Indeed, he recently told me that he undertook the service-learning activities partly as a calculated strategy to help better position himself for a job at a liberal arts college. As the essay by Dwyer points out, there is also a place in the mathematics department at a large research university for faculty with a passion for educational issues.

Author's Experience

In working with our service-learning community partners, it is important to realize that there is much we in higher education can learn from our community partners and that the interactions will usually be reciprocal with each side contributing and learning from the other. For example, although the college based math reforms efforts of the past decade have made mathematics faculty more aware of issues of pedagogy, our K–12 colleagues are often well ahead of us on this score.

I have improved my own college level teaching by adapting the innovative approaches to teaching that my children's elementary school teachers use as part of the Everyday Mathematics curriculum [5]. Their teaching has challenged me to think about the value of connecting the abstract math I teach with the real world and has led me to realize how valuable these connections can be in elevating students' interest in mathematics. Walking the halls of my children's school, I am constantly inspired by the students' wonderfully creative math projects that are on display. As an outgrowth of these reflections, my math colleagues and I instituted a poster session component to our multivariable calculus course. The students' assignment is to make a poster dealing with any topic of interest to them that is in any way connected to the material we have studied in the course. Poster is broadly defined and has included dance performances, songs, cheerleading routines, sculpture displays, and novellas. Topics included architecture, art, seashells, food shapes, and maps, as well as more traditional topics such as Keplerian orbits and quantum mechanics. The top posters get put on display in the math department hallway. Although I was nervous about trusting my students to successfully take on such an open-ended assignment, I was thrilled to see how well they responded, the creativity and initiative they showed, and how much of themselves they were able to bring to the project. The posters helped the students appreciate how smart and talented their classmates were and what a valuable resource they could be for each other.

As the relationship between partners strengthens and trust develops over time, participants should be on the lookout for opportunities for heightened levels of interaction. For example, a service-learning activity that initially involved service-learning students tutoring in a pre-college teacher's class might evolve so that the host teacher works with the service-learning students to jointly craft an activity-based lesson that will be incorporated directly into the classroom curriculum. Or pre-college teachers could be invited to be guest speakers in the college pre-service course, sharing their expertise on how the theoretical issues being discussed play out in practice. Service-learning host teachers are candidates to become mentors or supervisors for student-teacher placements. In such ways, the reciprocal nature of the relationship can be continually enhanced with the pre-college teacher becoming a valued colleague.

For some mathematics faculty, the prospect of adding a service-learning component to a course is a daunting challenge since disciplinary faculty have received little or no formal training in such non-content based pedagogical activities. Fortunately, education faculty are well situated to help math faculty think about how to conceptualize and implement such activities. In some institutions, there may be historical divisions between the mathematics and education faculty, with the latter being viewed as the poor relations. Service-learning provides an opportunity to build closer relations between the two groups as education faculty share their expertise to assist their disciplinary colleagues. I would expect that education faculty would be receptive to requests for such collaboration. A thoughtful department chair or dean could help to get the ball rolling in these contacts. Once the two parties start engaging in dialogue, they will likely find areas of common interest that will make the partnership self-sustaining.

Service-Learning Course on Changing Pedagogies

As an illustration of how some of the issues around service-learning play out in practice, I'll discuss my own experience in developing and implementing a service-learning course (called Praxis courses at my institution) and some of the unpredictable directions in which it has taken me. My involvement started with being invited to be a faculty representative on the service-learning oversight board that was developing guidelines for service-learning courses. Although I was initially unenthusiastic ("yet another thing to do"), I soon became a supporter of the service-learning program and started thinking of developing my own service-learning course around the theme of "Changing Pedagogies in Math and Science Education." The primary clientele for this course would be the large number of math and science majors who are not involved with our teacher education program; the secondary clientele would be the much smaller number of students who are preparing for careers in pre-college math and science education.

In my math teaching, I have used various non-traditional approaches (cooperative learning, computer labs, and student projects). I also started to learn and become excited about the developments in pre-college math reform, the Everyday Mathematics program at the elementary level that I mentioned previously, and the Interactive Mathematics

Program (IMP) that my school district began implementing at the high school level. These NSF-sponsored curricula are heavily inquiry-based and use challenging problems that make connections to the real world. A math major at my institution was in one of the first cohorts of high school students to follow the IMP curriculum. I was impressed by her independent and self-reliant learning style and wondered how much of those skills could be attributed to her IMP experience.

In designing the "Changing Pedagogies" courses, I aimed to expose our math and science majors to these new pedagogical approaches, with the goals of (i) interesting more of them in becoming teachers, (ii) giving the students preparing for careers in secondary school math and science education experience with these new approaches, and (iii) getting our graduate school bound majors to think about pedagogy in preparation for their future careers as professors. Although I felt knowledgeable about issues related to college level mathematics teaching, I did not feel I knew enough about pre-college math education or very much at all about science education to be able to develop and teach such a course on my own. We have a small education program at my institution that does not have anyone specializing in math or science education, but through my network of math colleagues, I linked up with a science education faculty member at a nearby institution who agreed to work with me to develop and teach the course. I was able to find funds from the college to pay her to co-teach the course, and we received a course development stipend from the college's service-learning grant to jointly design the course over the summer.

The course had a weekly lecture series and reflection session on issues of pedagogy. The students each had a placement, of four hours per week, with a host teacher who was involved in some aspect of pedagogical change in math or science. We invited the host teachers to attend the weekly lecture series associated with the course, both as audience participants and as speakers on panels. Having practicing teachers participate in the dialogue enriched and deepened the discourse as well as showed our students that we respected and valued the expertise of our pre-college teacher colleagues. The teachers who participated in the lecture series received continuing education credits which helped them fulfill state requirements.

Through our collaboration, I learned about issues involving field sites, student placements, reflection papers, and assessments using rubrics rather than tests. We plan to co-teach the course again, but the longer term goal is to develop in-house expertise among other math, science, and education faculty at my institution so that we can partner among ourselves in teaching the course.

Shortly after our course started, the Math Science Partnership program (NSF-02-061) was announced. This program aims to improve pre-college math and science education by building partnerships between school districts and institutions of higher education. My science education colleague and I became involved in efforts to organize a regional partnership, and we eventually became two of the co-principal investigators for the partnership, which recently received a five year, $12.5 million grant from the NSF.

Conclusions

To wrap up our discussion, let us look at future directions in which service-learning in mathematics might move. At present, service-learning acts as a way to improve undergraduates' understanding and engagement with mathematics and mathematics education. One could consider service-learning as a topic in its own right and provide explicit training in it so that future teachers can better institute service-learning in their own teaching. Students and faculty could work with pre-college teachers to help develop service-learning activities for K–12 students; in particular helping the teachers identify opportunities to include math content in these service-learning activities. For example, if the class is going to be measuring pollution in a nearby stream, the college math students could help identify and develop appropriate mathematical materials. As a follow up to this first service-learning activity, a later class of college students could work with the teacher as he or she implements the service-learning activity and help the younger students with the math activities. McDowell gives a wonderful example of this type of service-learning activity: her pre-service teachers developed math based activities for a school based community garden project.

There is much work still to be done in devising appropriate and effective models of service-learning in disciplinary courses, particularly advanced courses. One could develop a service-learning loop to tackle this problem. As the service-learning component of a Secondary Math Methods course, or perhaps an education course that includes a focus on service-learning, students could work with math faculty to develop proposals for service-learning activities in math content courses. The culminating experience could involve the students presenting their proposals to the math department. Once service-learning opportunities are identified, a future service-learning project could be to help the discipline faculty implement the new service-learning projects. As faculty examine the issue of quantitative literacy courses, they might consider including a service-learning component (Zang) as a way of making the material more relevant and thereby improving student interest and performance. There is also a need for more research on the effectiveness of service-learning in improving student learning and in changing student attitudes about mathematics. As the essays by Zang and Ridlon show, one can partner with a statistics colleague to undertake such research and can link this research to a service-learning activity for the statistics course.

As a way to spice up one's teaching so that both instructor and student benefit, service-learning has much to offer. I encourage you to give it a try.

Acknowledgements

I have been inspired in my service-learning activities by the social activism of Freya Von Moltke and my great uncle Eugene Rosenstock-Huessy, a pioneer in youth service and service-learning whose work at Dartmouth College was instrumental in creating the Camp William James project [3]. I thank Hillary Aisenstein, Nell Anderson, Michelle Francl, Jennifer Nichols, Deborah Pomeroy, Catherine Roberts, and Debra Rubin for helpful input as I was preparing this essay and Charles Hadlock for providing me with the opportunity to contribute to this discussion.

References

1. National Council of Teachers of Mathematics, *Principles and Standards for School Mathematics*, National Council of Teachers of Mathematics, Reston, Virgina, 2000.

2. J.D. Bransford et al., eds., *How People Learn*, National Academy Press, Washington, D.C., 2000.

3. Jack J. Preiss, *Camp William James*, Argo Books, Norwich, Vermont, 1978.

4. Lynn Arthur Steen, *Mathematics and Democracy: The Case for Quantitative Literacy*, The Mathematical Association of America, 2001.

5. University of Chicago School Mathematics Project, *Everyday Mathematics*, Everyday Learning Corp.

About the Author

Victor J. Donnay is Professor of Mathematics, and former Department Chair, at Bryn Mawr College. He is currently Co-Principal Investigator of the Math Science Partnership of Greater Philadelphia. He received his PhD from NYU's Courant Institute and was an Instructor at Princeton University. In addition to his research on chaotic dynamical systems, Professor Donnay and his Bryn Mawr students use computer graphics to bring the beauty and excitement of modern mathematics to the general public. Their work has been on display at the Maryland Science Center as part of the exhibit "Beyond Numbers."

4.2

Technology-College Algebra Service-Learning Project: "Surviving College Algebra"

Dana Craig

University of Central Oklahoma

Abstract. This section describes a project in which secondary mathematics education students developed a supplemental program for college algebra students at the same university. The focus was on the use of graphing calculators, especially for applied problems. This provided pedagogical experience for the education students while at the same time assisting the mathematics department to integrate calculators more thoroughly into the algebra course.

Introduction

The technology-college algebra service-learning project was designed to provide active participation in a teaching and learning setting to pre-service mathematics educators while at the same time addressing the needs of college algebra students at the University of Central Oklahoma. The pre-service teachers introduced real life applications involving the graphing calculator to college algebra students in a project that they titled "Surviving College Algebra."

The Mathematical Association of America has been promoting the integration of technology into mathematics courses for many years [1]. This service-learning project involved the use of technology in the form of graphing calculators. Some college algebra students are already proficient with graphing calculators while others have never had any experience with them. Mastering the graphing calculator is a priority learning target for the methods students because they need to be prepared to use them when teaching. By working with the calculators in a teaching role, these students can enhance their own mastery of the technology and of applied mathematics, and at the same time contribute to the needs of the algebra course.

The service-learning 'community partners' in this case included all sections of the college algebra course at the University of Central Oklahoma. The class and I hoped to provide these algebra students with an opportunity to understand their course subject matter from a new perspective by including more emphasis on technology and applications.

A Symbiotic Need for the Service-Learning Project

There are currently twenty-seven majors at the University of Central Oklahoma that require college algebra for the completion of the degree. Most students who need this course enroll in it during the first semester of their freshman year. Facing the challenges of adjusting to university life, the pains of math anxiety, and the differences in their high school math preparation, many of these students drop out of or fail college algebra in their first attempt. The university was thus eager to identify additional support for students enrolled in college algebra. Moreover, a new textbook had been adopted by the mathematics department that required the use of graphing calculators in assignments. We decided that a valuable project for the methods students would be to develop and implement problem-solving exercises that would integrate the use of such calculators. This project would be consistent with the goals of the National Council of Teachers of Mathematics, which encourage the use of real world applications to motivate conceptual understanding [2]. Thus, the service-learning project served the learning needs of both the mathematics education majors and also the college algebra students

The Methods of Teaching Secondary Mathematics class is the final course to be completed by pre-service mathematics educators before they begin student teaching. It is the only required course in which undergraduate mathematics education majors learn the pedagogical aspects of teaching and learning specifically related to mathematics, as is recommended by the National Council of Teachers of Mathematics, the Mathematical Association of America, and considerable current research. One section of the course is offered each fall with approximately twelve to fifteen students. One of the goals of the service-learning project was to prepare students who were enrolled in this course for the complexities and challenges involved in creating a curriculum that fulfills the needs of a variety of learners. I wanted them to have metacognitive discussions about their teaching experience and synthesize their teaching methods.

Many students had, of course, succeeded in becoming effective educators following this course as it had existed before, but I wanted to enhance their learning experience in the areas of technology and applications before they arrived in front of a high school classroom. The service-learning project complemented the lecture presentation in the classroom itself, while the students also developed greater creativity and self-confidence through developing applied and technology-related material.

The Technology-College Algebra Service Learning Project

The technology-college algebra service-learning project was an experimental venture combining the use of service-learning and graphing calculators to enhance the study of college algebra. A pre and post survey were administered to the students in the methods course concerning their opinions on the project. The specific focus of the project was on using graphing calculators to help explain mathematical concepts conceptually and to motivate students in connection with applications. The service-learning activity comprised twenty percent of the grade in the course. The scope of reflection papers and classroom discussion included metacognitive thought, problem solving, mathematical content, pedagogy, and the actual implementation of the service-learning project.

The pre-service teachers at the University of Central Oklahoma who enrolled in Methods of Teaching Secondary Mathematics were responsible for designing and implementing their service-learning project. Several became liaisons for the project, asking permission from college algebra professors to speak to their classes to advertise the special sessions they were offering. Fliers were placed throughout the classroom buildings publicizing the times and locations of service-learning sessions for the college algebra students, and times and dates were even written with sidewalk chalk at the entrances to the mathematics building. The mathematics department was supportive of the students' efforts and encouraged college algebra students to participate in the project, and, of course, coordination with the department was important. For example, algebra professors were surveyed to determine test dates so as to better prepare for the number of students that might want to attend the special service-learning sessions, and the names of students who attended were provided to professors upon request. Two professors requested this information, and one professor offered extra credit to his students who participated.

The methods students wrote five reflection papers concerning their perceptions of their experiences during the course. At the beginning, the papers contained quantitative, data-driven facts such as the number of students that attended, the amount of time spent and the number of students knowledgeable about graphing calculators. However, towards the end of the project, the methods students' papers evolved into insightful commentaries concerning the experience of teaching, thus demonstrating a new found awareness of the challenges of becoming an educator.

One of the most difficult obstacles in organizing the service-learning project involved scheduling and location. Because the project was extracurricular, many students had difficulty arranging time availability, given the constraints of part-time jobs and other classes. After the dates and times were established, difficulties arose in finding an available classroom. The methods students found after the first month of this project that a set time and place each week were easier for students to follow than a flexible weekly schedule. Students could then plan accordingly with their other extracurricular commitments.

Impact on the Education Students

The pre-service secondary mathematics teachers were able to experience first-hand the difficulties of teaching mathematical concepts and the use of graphing calculators to students with different levels of mathematical understanding. The pre-service teachers decided as a group on the approach that would be taken for each lesson. In general, for two hours a week for ten weeks, two students would present a real life application incorporating the use of technology. Each lesson examined a topic that corresponded to the appropriate algebra lesson from the department syllabus. The pre-service teachers reflected on and reevaluated weekly the progress of the project by surveying the college algebra students' needs and reactions. The college algebra students often wrote quite positive comments about the service-learning project. The pre-service teachers adjusted their methods and strategies accordingly to aid the students' technological and mathematical understanding. The hand-held graphing calculator proved to be a motivator for solving problems in a way that would allow the algebra students to concentrate on understanding the underlying mathematical concepts.

The methods students generally reported that they had developed a much deeper appreciation for the scope of skills needed in implementing curricular initiatives and special learning programs. They also benefited from the process of discussing how the algebra students actually learned in the course of the project. The service-learning project required a substantial investment by the students in preparation and time. Furthermore, the students were challenged to be creative in applying mathematics to the real world, and they were faced with the challenges of accommodating the needs of a range of different types of learners.

Towards the end of this semester-long project, the methods students began to connect their experience in this exercise with the theoretical content of their education program, and they seemed to be developing a greater sense of their career path as future professional educators. The actual experience of teaching during their service-learning project allowed them to gain metacognitive and in-depth insights into the challenges of effective teaching both as they observed their classmates and experienced for themselves the obstacles involved in applying mathematics to real life situations with technology.

Impact on the College Algebra Students

This service-learning project helped the university implement the use of the graphing calculator in college algebra, while providing additional resources for the college algebra students. The college algebra students also benefited from the additional curricular materials that included both technology and real life problem solving situations. The real life applications proved to be more accessible using the graphing calculator as a tool for conceptual explanation of problems. The math education majors involved in the project offered a range of teaching methods to meet the specific needs of their students, and the use of graphing calculators provided a positive environment for pre-service teachers to practice alternative techniques with college algebra students. Thus, the college algebra students' need for proficiency with graphing calculators and their struggles in completing college algebra requirements seemed to be aided by the pre-service math educators through this program.

Participant Assessments

The results of a survey confirmed that the pre-service teachers had a favorable reaction to their experiences in this project. Ninety-two percent of the students in the methods course responded to the post survey agreeing that the work done through this course benefited the university community. The results of the poll also showed evidence that methods students felt they had been effective in achieving their pedagogical objectives. Additionally, seventy-seven percent agreed that combining service work in the community with course work at the university should be practiced more. In spite of the occasional frustrations encountered in the project, the hard work and effort required, and the time-consuming nature of the assignment, the majority of the students reported that they not only benefited but also enjoyed this experience.

Though the majority of the students expressed positive statements about the project, I believe their actual individual levels of achievement varied widely and directly corresponded to their commitment and desire to teach middle school and high school mathematics. The students who wrote the most thoughtful reflections in their papers from this project were the ones who repeatedly expressed their desire and passion to teach mathematics and were already actively in the process of seeking teaching positions. The methods students who were unsure of their career plans, despite their planned student teaching and education degrees, still demonstrated good efforts, but they seemed less interested in the intrinsic rewards provided by the opportunity and more interested in their grades.

At the conclusion of the project, the students had a greater understanding of the multiple components of good teaching. My evaluation of the methods students' oral and written reflections throughout the service-learning project, along with the college algebra students' survey, all clearly and consistently conveyed that everyone involved was exposed to a highly effective and different learning environment with the technology-college algebra service-learning project.

References

1. Mathematical Association of America, *CBMS Mathematical Education of Teachers Project,* Washington, DC, 2000.
2. National Council of Teachers of Mathematics, *Principles and Standards for School Mathematics*, Virginia, 2000.

About the Author

Dana Craig is the Assessment Coordinator for the College of Education at the University of Central Oklahoma. She previously was an Assistant Professor of Mathematics and Statistics, where the service-learning project discussed in this section was initiated. She has also taught at the secondary level.

4.3

K–12 Math Tutoring as a Service-Learning Experience for Elementary Education Students

Jerry F. Dwyer
Texas Tech University

Abstract. University students in mathematics classes for elementary education attend local K–12 schools to assist in tutoring activities. This enhances the learning in their university class and they are awarded class credit. Procedures for teacher and student selection and enrollment are described in this chapter. Teacher feedback and student essays are discussed. Some suggestions for developing a service-learning program are offered, and constraints are also discussed.

Introduction

Service-learning in the University of Tennessee mathematics department, where this project took place, is defined as an activity in the community, performed by students, with the objective of enhancing their own learning, while providing a service to the community. This activity is considered part of a course and is assigned a grade in the same way as any other course component such as homework or other projects. One service-learning option is to perform teaching and tutoring duties in local elementary and middle school classrooms. The purpose of this chapter is to describe our experiences in setting up and implementing such a service-learning program. I will begin with an outline of activities and then describe those activities in more detail. Some of the description follows from my role as a director of the program and some relates to my personal experiences as a course instructor.

The majority of students who take a service-learning option are enrolled in mathematics courses designed for pre-service elementary education teachers. This option is an especially appropriate assignment for students taking such classes as it helps them discern whether they really want to teach and it also allows them to see the relevance of their curriculum in the actual classroom. The service-learning opportunity usually occurs in the freshman or sophomore years, before the students have a chance to participate in the regular pre-service program organized by the college of education. Students at this level are taking courses in elementary number theory and geometry. A few sections of mathematics or statistics classes for liberal arts students have offered service-learning, but relatively few students participate. For that reason I will focus this discussion on the elementary education specialists.

Service-learning is only one project option in these courses, and some students do not elect to take it. Over the last few years approximately half of the students took this option, resulting in about 25 students performing service-learning activities each semester. Other students do projects such as the investigation of number properties, reports on web searches of a mathematical topic, or design of a lesson plan. These projects usually require less of a time commitment than the service-learning option.

From a management perspective, the key components of our service-learning program may be summarized as follows:

1. Teachers from local K–12 schools are contacted to determine their interest.
2. Students from university classes are selected.
3. Students attend a K–12 classroom for one hour per week for a minimum of eight weeks.
4. Students tutor small groups, grade papers, observe teachers, work one on one with children, and participate in any other activity that is of benefit to the teacher and enhances their learning of the material in their university course.
5. Teachers return an evaluation form.
6. Students write a reflective essay on their experiences.
7. Classroom performance and essay are graded; this is usually ten percent of the class grade.

The sections below expand upon these components, and at the end I provide further discussion of the program in general.

Teacher Selection

The initial step in setting up the service-learning program is to contact teachers to see if they are interested in having a student in their classroom. In the first year of a new program this is a major step, but in subsequent years it becomes easier as more teachers become familiar with the program. Word spreads of the positive impact of the volunteers and demand builds for their services.

Some university instructors are not comfortable approaching teachers in the schools; it takes some of the abilities of a salesperson to promote the initiative. There are some proven strategies to assist in the process. One is to contact the local school district mathematics specialist and ask for a few minutes to speak at a workshop or in-service class for teachers. There will usually be plenty of interest in having volunteers who will help for one hour per week in their classrooms. This is also a good opportunity to outline the objectives of service-learning to K–12 teachers. They need to be made aware of the focus on learning as well as service. Holding a teacher workshop before the semester begins has proven to be a good way to meet teachers and describe the program to them.

Another good idea is to ask faculty colleagues for the names of teachers whom they know personally. Colleagues in other related disciplines may also be good contacts as they may already have some programs with local schools. The office of the dean of the college may also have a list of schools or organizations that seek volunteers. Opportunities for tutoring may even exist outside of the regular school setting, such as special education centers and homes for disadvantaged youth. Social events outside the classroom are also occasions to meet teachers. Every teacher is a potential participant in the program. I also call school principals and offer to present a math enrichment activity at their schools. When I am there I describe service-learning as one of the ways that our math department can assist in the classrooms.

I have found that after one or two years there are more teachers interested than we can accommodate. Quite often our participating students know teachers from their own K–12 days and approach them so that they can work in their own old classrooms. It is important in all approaches to teachers that we treat them as equals on a professional level. We can offer assistance in their classrooms and they can enrich the experience of our students. It is a mutually satisfying arrangement.

Student Selection

The second step in setting up a program is to identify students from the university class who are likely to succeed in their volunteer assignment. Some students may not be good ambassadors for the college. This is a reason to make service-learning an option rather than a requirement. There is also a reason to wait a few weeks after the beginning of class before allowing students to visit the schools. If a student is late for class, behind on homework or doesn't seem to work well with others, then it may not be appropriate to have that student representing the class in the outside community. The credibility of the program is very important and high standards must be maintained to ensure that teachers will have a good experience and want to continue to participate in subsequent semesters.

A course coordinator with responsibility for a number of instructors needs to meet with those instructors before class begins so that the objectives of service-learning can be well defined and transmitted. Further meetings should take place on a regular basis throughout the semester. Many faculty colleagues are not at all aware of what service-learning really means—some think it is just service, some think it is a full course in its own right, some think it is not for a grade, etc. It is important that these misconceptions be cleared up early on. It is also a good idea for the coordinator to visit individual K–12 classrooms occasionally to check on the status of each student's placement and to remind students of the deadlines. Students are notorious for not reading guidelines, and some will approach the instructor at mid semester wondering when to begin.

Student Enrollment

The next step is to formally enroll each participating student. The University of Tennessee requires that students purchase liability insurance or sign a waiver from such insurance. We work as a team on this. I am the course coordinator and I prepare the guidelines for students and teachers. I ask the other instructors (usually two or three) to distribute the guidelines and inform their students. I explain the program carefully to any new instructor. This is all done locally within the elementary education courses and does not require departmental approval. Nevertheless, the department chair is highly supportive of outreach activities and I have discussed (and promoted) service-learning with him. I have a dedicated outreach role and the coordination of these programs is seen as part of my regular service/research load (Conway [3]; Dwyer [5]). The student placement stage can often involve a substantial time commitment on the part of an individual instructor. I have approached the outreach office of the college of arts and sciences for some formal assistance and they have provided a graduate student to help with enrollment forms, insurance waivers, and placement in schools.

When a student enrolls we talk to each one individually about the expectations of their service. We stress the need to be responsible in attending the K–12 class regularly and punctually, and in notifying the teacher in a timely manner if scheduling conflicts occur. We outline the possible activities that they may participate in, such as small group tutoring or full class presentations. The students are given a form with guidelines (sample in Appendix I) for the teacher suggesting these same activities. It is important that the teachers be aware of the educational role of service-learning. The need to be of service and at the same time to enhance student learning is emphasized. When students are placed they must notify the departmental office of the location and time of their volunteer assignment. They are representing the university and performing formal course work; we need to have a record of their whereabouts.

When participants are from a particular elementary mathematics class at the university it is usual to have them work in elementary schools. This makes it easier to find activities that are directly relevant to their university curriculum. Some middle school assignments also work in this regard. Almost all of the schools are within a ten-mile radius of the university and most students have their own transportation. On rare occasions when a student doesn't have a car, we can arrange some car pooling or assign that student to a school that can be easily reached on foot or by bus.

It is important that the students be aware of the different roles of the faculty course instructor who is grading the service-learning essays and the graduate assistant who is coordinating placement. Students have to be constantly reminded of deadlines. We have one faculty member who grades all essays, and this is a major time commitment near the end of the semester. It is essential that all work be submitted one or two weeks prior to the end of classes.

Grading

We usually award ten points for the service-learning project, which amounts to ten percent of the total class grade. If the student volunteers for at least eight class periods then she/he is automatically awarded four points. If the teacher feedback is positive then another three points are awarded. For the essay the student is awarded a point for describing the activities (e.g., a journal type account), a point for observation, and a point for connecting the volunteer work to the university class material. Observations may include new insights obtained into teaching, student attitude to teaching, or how children learn.

The amount of points for the reflection may seem low, but there are only ten points available and students should get several of those points simply for attending. This keeps the grade consistent with other students who do not take the service-learning option and who can obtain high points for routine work. Otherwise the service-learning students may feel that they are not being treated fairly. In general most students obtain about eight points. They all seem to do a wonderful job in the classroom but they don't always offer adequate reflection, as I explain below.

Student Reflections

The student essay at the end of the semester is extremely important. It helps us to see what they have done and what effect it has had on their education. In our particular situation, student comments on the experience were very positive. They said that the volunteering made them feel good about themselves. They were able to help others, gain a sense of community, and were able to act as role models for the children, especially in some schools with large numbers of minority students. Often they remained at the school for much longer than the required period. Some said that they would return as volunteers the following semester. Quite often they helped with reading and other subjects as well as assisting with the math classes. They also bonded with many of the children and had a sense of sadness when they were leaving.

The most positive endorsement came from those students who said that they would change their major to education as a result of this project. Many of them were happy to volunteer for the experience independent of the course credit offered. The project was also instructive for a small number of students who were able to decide that they were not suited to this role. Others were able to determine which grade levels they were most interested in teaching.

Many students felt that their own learning skills had improved. Some commented how they had to revisit many of their computational skills and found that to be challenging without the use of a calculator. This is a clear case of where the university students' math learning is enhanced by their tutoring experience. They learned how to be more patient with others and with themselves. Working with children taught them about the need for persistence in solving problems and they could translate this to their own learning. They gained a much greater respect for the teachers and the demands of teaching, and they also gained a greater understanding of diversity and the challenges facing children from disadvantaged backgrounds.

One reservation expressed by some students was that they weren't sure of the exact obligations or requirements once they were in the school. In the essays there was very little comment on the direct connection between the volunteer teaching and the learning of the material in their university class. For example, if students are tutoring children on different ways to do long division, then that should clarify their understanding of the long division algorithms they are studying in their elementary math course here on campus. The participants don't always make the connection or write about it in their essays. Conversely they do see how their university course helps them in their tutoring, but that is not the focus. It is a worthy objective but the major intent of service-learning is precisely the other way around—the service must enhance their learning. Some of these misunderstandings may be due to insufficient guidelines on our part. The individual instructors don't always have the same commitment to service-learning or awareness of the full impact of the program, and the service-learning coordinator needs to give clearer directions at times.

Teacher Evaluations

Teacher evaluations were solicited through a feedback form (sample in Appendix II) with a number of questions. They were asked to outline the activities of the service-learning student, the impact of the student in the classroom, the behavior/attitude of the student, and an overall impression of the program with suggestions for its improvement.

Teachers were generally pleased with the program. They valued the extra help in the classroom and appreciated the enthusiasm and professionalism of the volunteers. They could see the positive effects on their students and asked that they be assigned new volunteers as the program continues. One interesting observation concerned the positive role model effect of female volunteers on young girls. In many areas women are still not accustomed to entering the scientific and technical disciplines, and it is very beneficial for girls to see young college women who are taking on these roles.

Some teachers were concerned that the project was not well defined; they weren't sure exactly what the students were supposed to be doing while in their classrooms. There were also some comments regarding volunteer dress codes and appropriate behavior. It would be helpful if the instructor could visit the school at least once in the semester to have a one on one conversation with the teacher. Such school visits would give the teachers a clearer understanding of service-learning, as teachers often assume that this is just another routine volunteer activity rather than a formal service-learning experience.

The teacher feedback has proved very useful. We have been able to modify our approach to facilitate their concerns. We now include a form for the teacher, which outlines the expectations for this type of experience. We have also been able to advise students more comprehensively on issues of responsibility in the classrooms. In this way we are truly building a partnership between the teachers and the university mathematics department.

Discussion

I believe strongly in the value of service-learning. There are great benefits to the community and to the participating students. However, we need to point out clearly that they are doing service and learning rather than just service. It seems that the culture of volunteering and service is well ingrained, but that there is a challenge to create awareness among students and teachers of the dual role of service and learning at the same time. We must also be aware of the need to make a genuine contribution to math education at the K–12 level, rather than just fulfilling a project requirement.

It is difficult to evaluate the effectiveness of a service-learning program. This in turn makes it difficult to publish in the educational literature, as formal assessment is often a necessity for those publications. Student surveys can be subjective, and students' experience in service-learning cannot be compared with others who did not take part. In some of our sections almost all students opted for the service-learning project, and there was no random control group remaining for comparison. Another issue is that students who do service-learning tend to be well motivated in the first place and may perform better in all other aspects of their course also. At this stage we are happy to have a program that creates a strong level of civic engagement and much community satisfaction, whether or not we can formally document major academic achievements. I would hope in the future to work more closely with education specialists so that we could design some formal assessments.

I estimate that as coordinator I spend about two hours per semester on each student who enrolls in the service learning program. That includes meeting the student for about 15 minutes per semester, making a classroom assignment, talking to the teacher, dealing with complaints, emails, etc, and grading the final essay. That amounts to about 40 hours extra work per semester if 20 students participate. The total time is not excessive and several of these tasks take less time as we become familiar with the teachers and the logistics of the program. However one must be aware that most of the dedicated time occurs during the first month of the semester and again in the last week, and careful planning has to take place to ensure available time at those critical stages.

There is also a challenge involved in coordinating the activities of the various participants in a service-learning program. These include the course instructors, university students, university administrators, K–12 teachers, and school officials. It is essential that all channels of communication be open; the program director requires some managerial skills to keep all parties functioning smoothly.

Assistance from the campus administration is essential if faculty members are to become involved in service-learning. In our case the college's outreach office has been most helpful as the graduate assistants have performed literature searches and kept us up to date on the latest developments and meetings. As a mathematics instructor, my primary interest is in teaching and doing mathematics. I am very glad to participate in beneficial educational programs such as service-learning but I am reluctant to spend much time reviewing the literature or writing too much about this. The average instructor simply does not have the time to organize and facilitate these activities. Incentives have to be provided in the form of grants and recognition in terms of promotion and tenure.

The literature on service-learning has expanded rapidly in recent years but there are still relatively few examples of service-learning in mathematics (prior to this volume). Let me mention a small number of articles that discuss how others are also pursuing this work. A good article by Duke [4] describes some projects in statistical classes and also comments on instructors' concerns. Conrad and Hedin [2] report that programs in which more advanced students tutor less advanced students resulted in higher mathematics achievement scores for both the tutors and those being tutored. Austin et al. [1] describe a program where students interview a person who described a real algebra problem being solved in the real world. The feedback from students suggested that the project helped them see how the subject matter can be used in everyday life.

A sample review of web sites reveals that many service-learning options in mathematics involve some kind of K–12 classroom tutoring. Several of these experiences are related to elementary teacher training courses. These are valuable services for both students and teachers but the learning experience is limited to students taking lower level algebra courses. There is a need to develop service-learning options for mathematics majors in more advanced courses. A small number of schools report activities in environmental modeling and data collection activities, and some of these initiatives are discussed elsewhere in this volume. Assistance in the use of spreadsheets for local non-profit organizations is another good option, and the development of software and artwork based on advanced mathematical ideas has also been reported. Much thought and innovation are needed to make progress in these more advanced areas.

Conclusion

I offer the following closing suggestions to emphasize what is described above.

- Start with a small group
- Talk to the dean or college office personnel
- Talk to K–12 teachers
- Identify good students and emphasize deadlines
- Talk to students during the semester
- Visit schools occasionally

I recently met a teacher who participated in our program and she commented that one volunteer from a couple of years ago continues to visit her classroom every week. That is an amazing testimony to the kind of long-term benefits and life changing potential that service-learning can produce. It encourages us to keep going when the administrative challenges seem daunting. This potential to make a difference in the lives of our students is very powerful and is one of the great rewards we experience as teachers.

Appendix I

Service learning expectations form

One of our students, _____ will serve as a volunteer in your classroom for a number of weeks as part of our service learning program in The University of Tennessee Mathematics Department. The student will obtain a grade in his/her UT class for the work done in your classroom.

The objective of service learning is to enhance student learning, while providing a service to the community. This gives our students an opportunity to understand the math concepts more fully and we hope that it is helpful to you and your students. This work is solely for the volunteer's UT Math class and is independent of any other volunteer or internship activity associated with the University.

The following activities are suggested:

Assist the teacher in explaining difficult material

Work with small groups of weaker students, who may need extra help

Work with small groups of advanced students who require extra activities such as math games

Make a presentation to the whole class

These are only suggestions. You are free, of course, if you wish, to choose other activities that would be helpful.

If you have any questions or comments I will be happy to assist you. Please contact me by e-mail at dwyer@math.utk.edu or by telephone at 974-4290.

The student is expected to attend your classroom at least eight times. At the end of that time we would greatly appreciate your feedback on the questionnaire on the back of this sheet. That will help the UT Math instructor to assign a grade to the volunteer and it will also help us to provide a better program to meet your needs in the classroom.

Thank you for participating in this program. We hope that it will be a great experience for you and your students.

Appendix II

Student evaluation form

Estimate the number of times the student attended your class and briefly describe the activities performed.

Was the student punctual, responsible, mature and cordial?

What was your overall impression of the impact of the student in your classroom?

We would like this service-learning program to be of benefit to you. We welcome your comments on how it could be improved.

Thank you very much for your time.

Please mail this form before Nov 30 to
Dr. Jerry Dwyer, Mathematics Department, University of Tennessee, Knoxville, TN 37996

References

1. S. S. Austin, C. L. Berceli, and S. Mathews. "When Will I Ever Use This Stuff Anyway?" *Mathematics Teacher*, 92 (9) (1999) 798–799.
2. D. Conrad and D. Hedin. "School-Based Community Service: What We Know from Research and Theory," *Phi Delta Kappan*, 72 (1991) 743–749.
3. J. B. Conway. "Reflections of a Department Head on Outreach Mathematics," *Notices of the AMS*, 48(10) (2001) 1169–1172.
4. J. I. Duke. "Service Learning: Taking Mathematics Into The Real World," *Mathematics Teacher*, 92 (9) (1999) 794–796.
5. Jerry F. Dwyer. "Reflections of an Outreach Mathematician," *Notices of the AMS*, 48(10) (2001) 1173–1175.

About the Author

Jerry F. Dwyer (PhD National University of Ireland) is an assistant professor in the mathematics and statistics department at Texas Tech University. The work reported in this article was carried out at the University of Tennessee Knoxville. He has taught a wide range of mathematics and engineering courses and has published mostly in the area of applied mathematics with applications in mechanics. In recent years he has organized numerous outreach and service-learning programs.

4.4

Service-Learning for Pre-service Teachers and Developmental Math Students

John F. Hamman
Anne Arundel Community College

Abstract. Service-learning has proved very successful at the community college level both for students in math classes for pre-service teachers and for students in developmental math classes. Students in both types of courses can choose to tutor at local schools or after-school programs to fulfill part of the course requirements. Tutoring younger students changes students' attitudes about mathematics and increases their motivation to learn mathematics, making the service-learning program truly meaningful to all participants.

Introduction

I have had success using service-learning in two different mathematics courses at Anne Arundel Community College: a two-semester sequence for pre-service teachers and a developmental mathematics course designed for students who need to master high school level mathematics so they can later enroll in college algebra or the equivalent. In both courses, learning is enhanced by student participation in mathematics tutoring at local schools and after-school programs. I use similar procedures to incorporate service-learning into the two courses, but the objectives and outcomes are quite different in the different settings. In this paper, I will break the process I use for incorporating service-learning into a class into the following five components: motivation for using service-learning; initiating the process in a class; assessing students' progress throughout the semester; reflecting with students on the experience at the end of the semester; and dealing with challenges that I have encountered.

Motivation for Using Service-Learning

Pre-service Teachers

One of the problems I have always faced in teaching math courses to pre-service teachers is getting the students to understand the depth of mathematical knowledge required to teach elementary school. Students often incorrectly assume that anyone who is capable of doing elementary school mathematics is capable of teaching elementary school mathematics. I want my students to have the opportunity to see the depth of mathematical questions asked by young children, and to see how misconceptions in the early grades can lead to serious misunderstanding in subsequent schooling. Additionally, I hope that my students will start to see mathematics as a creative process instead of as a series of memorized procedures and computations with one right answer.

All of these students will spend a great deal of time in the classroom before they become certified teachers, but the vast majority of that time will be long after they have left my class. I want them to have classroom experience while I am there to help them reflect on it and find ways to improve their understanding of the mathematics involved. Additionally, since one of the missions of my community college is to be an active part of the larger community, I want to establish meaningful connections between my mathematics students and the surrounding community. Service-learning is a natural choice that meets all these needs.

While I would like to make service-learning a requirement for all students in the course, I have too many students with families, full-time jobs and other obligations for this to be realistic. Therefore, I give students the option either of doing the service-learning, which requires 12 hours of tutoring over the course of the semester plus a daily journal and a final reflection paper, or of writing a research paper, which is designed to require about the same time commitment. I emphasize how important and useful I think the service-learning is. I also try to make the research paper sound as unappealing as possible. Usually about 80 to 90 percent of students choose the service-learning option. Either option is worth 10 percent of the final grade. The service-learning is graded solely upon completion of the three requirements mentioned above.

Developmental Math Students

For my developmental math students, my goals are to change the way in which they perceive mathematics and to help them learn study skills. Many of them believe that they cannot do any math; furthermore, some have never learned how to study mathematics. Service-learning helps students both feel better about their mathematical abilities and learn more effective study habits. When the students become the experts in the eyes of children, the students' perceptions of their own capabilities improve dramatically. Also, seeing how different study habits in children affect the children's grades teaches my students a lesson in how to be more effective in their own study time. The combined effect of these lessons on my students has made some of my weakest students excel and even enjoy mathematics by the end of the semester.

I use service-learning in my developmental courses as an option (although it is a strongly encouraged option). It is the only form of extra credit my students can earn all semester. If they complete all three requirements of service-learning (12 hours of tutoring, a daily journal, and a reflection paper), I replace their lowest exam score with an A. Even with this generous offer, less than 10 percent of my developmental students sign up for service-learning. Figure 4.4.1 shows one of my developmental students in the classroom.

Figure 4.4.1. Developmental math student at work on her project.

Initiating the Activity

On the first day of class, I describe what service-learning is and why I think it is important for them to consider. The following week I have a member of the service-learning office visit my class to further explain the benefits as well as the procedures for participation.

Students are given a booklet of agencies (compiled and written by our service-learning office) and the application form. Within the first month of classes, they need to identify an agency in the booklet that offers mathematics tutoring to public school students. Our list currently includes nine after-school programs, four elementary schools, and two special education services. Students can pick any grade level with which they feel comfortable. Each agency listing has a brief description of what the agency does, the hours they are open, how many volunteers they need, and contact information.

Often the hardest tasks for my students are deciding to participate and making a phone call to a complete stranger. The booklet has a list of opening lines and sample scripts to guide nervous students. They need to set up a time to meet with the agency and discuss how they will spend the 12 hours that I require of them. At this initial meeting, the student and the agency representative fill out an application saying what is expected, and they both sign it. The student then brings this form to me so I know what to expect and can see what level of mathematics is involved. If I approve the choice, I sign the form and file it with the service-learning office. If I do not approve it, I discuss my concerns with the student and he or she either finds a different agency or makes some changes to the initial idea.

Ongoing Assessment

Every three to four weeks throughout the semester, I take a few minutes of class time to discuss service-learning with my pre-service teachers. I want my students to share not only what they have learned but also their frustrations and questions. It helps students to hear that other students have been stumped by math questions, had to handle disruptive children, and faced other similar problems. The most effective learning, however, comes not from these discussions but from sudden outbursts in class when we are talking about models for fraction addition or a similar topic. Students will spontaneously share that they were trying to explain adding fractions to their students but didn't know what to say. The distinction between knowing how to do mathematics and how to explain mathematics becomes evident to all my service-learning students. It is amazing to see how motivated the students become. The mathematics suddenly becomes part of their immediate world, and they want to really understand it.

Unlike my pre-service teacher classes, so few of my students in developmental mathematics are participating in service-learning that I spend very little class time talking about service-learning during the semester. Instead, I talk to the students involved on a one-on-one basis. I want to find out if the student is learning during this process and to see if there are any challenges that I can help with.

I require all of my students, whether pre-service or developmental, to keep a daily journal of their service-learning experience. Every day, they need to keep track of what they did at the agency and their reactions to the day's events. The reason for keeping a journal is two-fold. First, the student needs a record to aid in writing the reflection paper. Second, I need a way to keep track of what various agencies are having my students do. When I collect the journals at the end of the semester, I make notes of which schools used my students effectively in their math classes and which simply had them sit in the back and watch math lectures. Likewise, I see which after-school programs had useful tutoring programs set up and which were acting more like daycare. This information helps me decide which agencies to recommend to students the following semester and which should be taken out of the service-learning booklet. For my pre-service teachers, I have a good sense of what each student is doing from our discussions throughout the semester. My developmental students, on the other hand, have given me little feedback. Therefore, I often ask my developmental students to hand in their journal halfway through the process so I can see what they are doing before their time is finished.

Additionally, each student has a daily log sheet that the teacher or supervisor signs at every visit. For the last day, the supervisor fills out a simple rubric asking about appropriate behavior, timeliness, and the like.

Reflections

The reflection process is an extremely important aspect of service-learning. The students write their reflection after all of their hours have been completed; it asks them to think about how their service-learning has helped them. Most importantly, it lets them think about what they have learned this semester both in and outside of class. It is important that my students be given the time and opportunity to really think about what they have done and learned. To help ensure this, we talk about the reflections in class several times before I give them the actual details of the assignment. I wait until several students have finished their hours before giving the class the description of what goes into their reflection paper, so no one does it early. Waiting to give out details allows us to discuss the reflection in class when it is most helpful to the students. I ask every student to hand in a draft of the reflection so I can read, make comments, and talk to the student before he or she writes the final draft. I almost never make comments about the grammar of this assignment. Instead, I talk about the content and how the things they learned apply to the course they are taking. This process of asking them direct questions about how they saw students learning and what processes were more effective than others is extremely beneficial for all of my students, but particularly for those in developmental math with poor study skills.

The reflection paper, which I require to be about 500 words, should answer at least three of the questions posed below.

How has your service-learning experience changed your ideas about:
- the importance or relevance of mathematics?
- the definition or nature of mathematics?
- your ability and interest in mathematics?
- effective methods to study and learn mathematics?
- your knowledge and understanding of mathematics?

How has your service-learning experience helped to:
- develop your general communication skills of reading, writing, and speaking?
- develop your specific communication skills in mathematics?
- clarify your ideas about the responsibilities of a citizen of your community?
- establish yourself as a contributing citizen of your community?
- clarify your personal objectives and your understanding of yourself?

Reactions from Pre-service Teachers

Perhaps the best way to show how my students assessed their service-learning experience is by giving you some samples from their reflections. The following are excerpts from the comments of five different pre-service teachers.

Student #1

Connecting with the children was amazing, and knowing that they were learning from me was almost spiritual. Doing the service learning has definitely changed my ability and interest in mathematics. Through service learning and [this class], I have learned that I am not as bad at math as I always say I am. I have taken a greater interest in it because I now see how important math really is.

Student #2

Before taking this class and completing my hours of service learning, I had a negative attitude about math. Now I tend to think more positively about math. When I was stuck, I would just give up, but after taking this class and doing the service learning I learned that it is going to be hard to get my students unstuck if I cannot get myself unstuck.

Student #3

The definition of mathematics that I had prior to doing the service learning was a series of steps used in order to complete an equation or problem in order to get a final and correct answer. My definition of math has changed. I think it is important to understand how a child thinks about a particular question in order to help that child understand the entire concept, because that in itself is what math is all about. The definition that I now have of mathematics is that math is problem solving.

Student #4

Being a teacher of math is not so much what you know and how much you can successfully compute in a given math topic, but rather how deeply or abstractly you understand the areas of math and how well you are able to transfer this knowledge to your students. Watching the light bulb "click on" in the facial expressions of some young individuals is a great feeling and a great sense of accomplishment for the child as well. Having learned this by first-hand experience, I have realized that mathematics is a subject that requires much knowledge and a true understanding of the concepts involved in order to be prepared to teach it to all types of student cognitive abilities.

Student #5

I admit that I wasn't thrilled when I heard that we had to complete service learning as part of our final grade. [But] through observing and assisting… I learned alternative ways to teach math from the way I was taught. I also realized that the way I was taught might not have been as effective as it should have been.

Reactions from Developmental Math Students

The following are samples from three different students in my developmental math classes.

Student #1

…math is far from being a good subject for me. As I sit here I see how my math strategy was wrong for learning math. As I sat and watched those children, they taught me that anything is possible in learning math. So when I say I can't do this, or math is impossible for me to learn; that's wrong, nothing is impossible. I am glad I decided to take the plunge and participate in service learning.

Student #2

Since doing service learning I have buckled down and actually started studying. Usually I blow things off for the last moment, but lately I feel that these students work hard to do math problems that come simply to me. So they gave me the initiative to start studying.

Student #3

The experience of this semester along with my participation as a tutor has boosted my confidence in my math ability tremendously. I was skeptical about tutoring a subject I perceived myself as being mediocre at but was willing to give it a try….

In my tutoring work my students repeatedly tell me that they don't want or need any help from me. They are very reticent to admit they don't know something much less outright ask for help. It almost seems as if they'd rather avoid math entirely than risk revealing their fear of math or any weakness in the subject. By learning about my own blocks and how to overcome them I've developed a more can-do attitude.

Challenges

I have had some challenges associated with introducing service-learning into my curriculum. The first challenge was to find agencies doing relevant mathematics that are willing to take students for a short time period. Luckily, Anne Arundel Community College has a service-learning office that has set up many sites for me. Still, we do not have sites that meet all students' needs. Some of the local schools will only take students if they commit for a much longer time. Additionally, some of the schools require hours of sexual harassment training and legal paperwork before a student can enter the classroom. Several students only have time available on the weekends, and it is very difficult to find tutoring opportunities for them. Therefore, unfortunately, some students do not participate in service-learning simply because we have not found an agency that is convenient enough for them.

Once the relationships with the agencies have been set up, the next challenge comes from the students. Most students are not pleased to hear that they are expected to spend hours doing mathematics outside of class in addition to their regular homework. I combat this problem in several ways. First, there is a special service-learning designation next to the class listing in the schedule, so that students have some advance notice of the fact that they will be expected to spend time outside of class doing class work. Second, at the beginning of the semester, I remind students that there is a college policy regarding the amount of time they are expected to spend on coursework outside of class, and I tell them that time spent on service-learning will fall within those parameters.

Finally, I designate several class periods throughout the semester as days set aside for students to do work associated with their service-learning. I do not assign any homework on those days, and I make a point of mentioning in class that students should use the time for their service-learning.

Lastly, I do not have the time to check out all the agencies myself. I have to rely on the service-learning office and student feedback to see if my students are getting what they should out of this experience. When I first tried service-learning, I had very limited discussions and no real feedback from the students until the end of the semester. I had a couple of students who ended up reshelving books at the library and supervising recess for much of their time. This experience taught me the importance of checking in with all service-learning students during their time at the agency. It also taught me the importance of having students go to the agency once before the service-learning begins to discuss what their role will be.

Conclusion

The ideas presented here can be used in a variety of math classes. The skills that my developmental mathematics students gain from this experience can benefit a student at any level of mathematics; even more important than the skills are the noticeable changes in attitude. Similarly, students in the mathematics courses for pre-service teachers gain not only a deeper understanding of the skills needed to teach mathematics effectively but also a deeper understanding of the mathematics itself. Students who participate in service-learning stop seeing mathematics as something that you either get or you don't and begin to understand the processes involved in thinking mathematically.

About the Author

John Hamman is an Assistant Professor of Mathematics at Anne Arundel Community College in Arnold, Maryland. One of his many duties on campus is to serve as a member of the service-learning advisory board. His research interests include logic and mathematics education. His non-mathematical interests include beekeeping, woodworking, and spending time with his wife, daughter, and new son.

4.5

America Counts: A Tutoring Program for the Twenty-first Century

Candice L. Ridlon
Brigham Young University

Abstract. A collaborative effort between a university and an urban city school district yielded valuable results. The university increased community recognition and received additional funds from the federal government. Pre-service teachers gained valuable classroom experience early in their careers, and all tutors, regardless of college major, felt a positive personal impact. Beyond simply improving children's mathematical skills, tutors became mathematical role models to the children they tutored, encouraging these at-risk students to raise their expectations and think about attending college themselves.

Introduction

When Towson University sent me to a meeting in May 2000 about a program called *America Counts*, I was surprised at the diversity of people attending this workshop. University administrators, financial aid officers, mathematics professors, curriculum materials advocates, foster grandparent clubs, and youth directors had all come together to learn about the program, a United States Department of Education (USDE) initiative. But as the 2-day workshop evolved, I discovered the reasons behind this diverse attendance. Loosely defined, *America Counts* is a program for improving mathematics curriculum and teaching practices through the coordination of federal, state, and local resources [1]. The initiative is multi-dimensional, and many partners must participate to implement it properly.

Having experience as a teacher, school volunteer, tutor, and mathematics education professor, I came away from the meeting with a personal vision of how my university could begin an effective mathematics service-learning project under the guidelines of *America Counts*. Thus Towson University's (TU's) tutoring program began in August 2000 as a cooperative effort with the Baltimore City Public School System. The school sites with the greatest academic need were identified, and then student and volunteer tutors were recruited, trained, and placed in these locations. We began with three sites, adding two in 2001 and two in 2002. The total of *America Counts* sites is currently seven, with many schools waiting.

Objectives and Goals

As Project Director for Towson University's *America Counts* program, I was free to target the school population of my choice. As TU is located near the city of Baltimore with its myriad urban educational and social problems, I chose to design a program for the specific needs of this community. However, *America Counts* could be equally effective with any group of students who are struggling academically. The objective of our program was to provide one-on-one tutoring and mentoring to low achieving, low socioeconomic level students, with the goal of preparing those students for admission to and success in post-secondary education.

Figure 4.5.1. One-on-one tutoring has been shown effective for improving student achievement.

Research has shown that one-on-one tutoring is effective in improving student performance. Because tutors work with individual students, they are able to customize an academic plan to remediate skill deficits. For example, they can re-teach mathematical concepts independent of the classroom curriculum. Furthermore, the lessons that non-certified tutors need for this kind of teaching are virtually free to participants who register with the *America Counts* initiative. These materials were designed by Math Partners [2] in collaboration with the USDE and the National Science Foundation. They are readily available through the Internet in the public domain, allowing for viable replication of the program at various sites.

The goal of preparing inner-city youth for further education was achieved by enriching their understanding of mathematics at an early age, while mentoring their interest in postsecondary education through continued positive exposure to college-age tutors. TU's *America Counts* program was designed with the following objectives:

1. To increase the percentage of children who pass the Maryland Functional Test (MFT). A passing score on the MFT is required before students attend regular ninth grade; failing scores and subsequent cynicism as students remain behind in middle school contribute significantly to the dropout rate in Baltimore schools.
2. To improve the mathematics skills of all students who receive tutoring services.
3. To give each tutored student a positive role model in the form of the tutor with whom he or she is matched. As students themselves, the tutors build individual relationships with their tutees while mentoring them in connection with their possible desire for and interest in postsecondary education.

4. To give children experience with appropriate technology needed for academic success (such as graphing calculators and geometry software).
5. To stage Parent Nights to promote an awareness of the importance of parent involvement in children's academic learning and to provide resources to encourage that involvement.
6. To provide professional development for teachers working with urban youth, especially in the areas of hands-on learning strategies and technology use.

Criteria for Schools and Students

All participating schools were screened to meet the selection criteria of Towson University's *America Counts* program. First, at least 90% of the students at the site must be on the free or reduced-price meal program. Second, a minimum of 50% of students must have failed the MFT, a percentage that does not meet state standards. Because students at these schools need considerable academic help and are from disadvantaged socioeconomic backgrounds, they represent the target population for *America Counts* aid. Without the services that a program like *America Counts* provides, their poor performance and lack of positive role models generally prevent them from considering higher education. They seldom acquire the resources to escape their inner city neighborhood, and the repeating cycle of poverty remains unbroken.

To maintain independence in implementing the program, each school developed its own criteria for identifying children chosen to receive tutoring services. Some sites selected students of any age who were nearest a passing MFT score, hoping to bolster them into success by the next testing date. Other schools chose students with the lowest MFT scores, trying to supplement the instruction of their crowded classrooms for those who needed it most. Still others concentrated exclusively on their oldest students, trying to get them through the MFT and into high school.

Collaboration with Others

Towson University was not the only university placing tutors in schools in the State of Maryland, but the TU program was the only *America Counts* program conducted as a collaborative partnership between participating schools and an institution of higher education. I worked directly with selected public school sites to set up a designated tutoring lab and trained the staff to support it. Each of the sites signed a contract with TU's Career Center, agreeing to provide a room for tutoring with specified hours of operation (their choice), as

Figure 4.5.2. University tutors became positive role models to youth when they developed one-on-one relationships with them.

well as an individual to supervise the lab. This supervisor agreed to identify children to be tutored, to verify each TU tutor's work hours, and to sign paid tutors' bi-weekly timesheets. Our program was unique in that it took advantage of available resources to blend service-learning and paid employment opportunities for our students.

The TU *America Counts* program was also conducted in collaboration with Greater Homewood Community Corporation, Inc., a non-profit organization with the charge of strengthening community organizations. Greater Homewood functioned as a "grants clearinghouse" for the Baltimore area, securing *AmeriCorps* personnel who worked as tutoring lab supervisors at three of the sites. All of the program partners recruited the services of additional community volunteers, including retired teachers, parents, and grandparents.

Although tutoring met the service-learning component of several psychology and business courses at Towson University, as well as the volunteerism requirement of the Honors College and the service-learning element of the Teacher Education Department, the Financial Aid Department was an important collaborator in shaping the success of the *America Counts* program. They allocated funds to financially needy students through an award called the Federal Work-Study (FWS). The financial aid department at any university, public or private, has a vested interest in supporting programs like *America Counts*, for the USDE requires that 7% of an institution's FWS budget be allocated to placing students in service-oriented community jobs. (Current talk in the political arena may raise this percentage to 13-15% or even higher.) Thus, it was in the university's and the community's best interest to encourage FWS students to become tutors in *America Counts*. There were acceptable ways to make this job attractive to students. Since the tutoring takes place off-campus, we paid tutors more than they would earn in other FWS jobs ($10/hour). We also awarded them 1 hour of pay for travel, entered on their timesheets, each day that they went out to work, and we paid for 1 hour of "teacher planning" for every 3 hours that they tutored (because proper lesson preparation takes time outside of the actual tutoring).

Because of Towson University's commitment to *America Counts*, the university also covered many expenses other than wages for tutors. This included cash contributions from the Department of Mathematics for photocopying, donation of materials like paper and classroom facilities, and my time spent in preparing training materials, recruiting, training, and placing tutors, and coordinating all activities. Although in the last two years a grant provided 3-hour class release time for me, in the first year the Department donated the faculty-release time. Towson University saw organized tutoring as strengthening an already strong community-service tradition.

Furthermore, the TU Marketing Department chose the *America Counts* program as a vehicle for community outreach. They purchased materials, donated Towson University logo merchandise for student incentives, and printed booklets for parents to use in helping their child learn math at home. In addition, they paid for *America Counts* T-shirts bearing the Towson logo for the tutors to wear at work (their "uniforms"). All of these donations managed to make the Towson University name a recognizable presence in the Baltimore schools and the community at large.

In addition, I secured $500 from the Exxon Mobil Foundation and $2000 in third party contributions in support of project activities. In 2001, the Maryland Higher Education

Commission (MHEC) funded a grant for $60,000 to expand *America Counts*. The following year (2002), they funded $50,000 for support of the program [3]. These funds hired additional tutors and purchased student incentives, software, calculators, games, workbooks, and training materials. They also provided incidentals, such as the program website, refreshments for the 8-hour training sessions, and an end-of-the-year picnic.

The Tutors: Recruitment and Training

University tutors came from every major and class rank in the university. We had everything from film majors to pre-law, from freshmen to graduate students. The main qualification for the job was a desire to work with kids. To recruit tutors, a campus-wide announcement was sent via e-mail at the beginning of each semester. Posters advertising the job were hung all over campus. Flyers touting the opportunity were distributed to faculty mailboxes and read in class. We set up a table in the Student Union at the Part-Time Job Fair. All recruitment efforts publicized a deadline for application and provided the training date.

Figure 4.5.3. Tutors came from every major and class rank—the main qualification was a desire to work with children.

Students were encouraged to first check with the Financial Aid Department to see if they qualified for Federal Work-Study—the surest way of getting paid. (The MHEC grant only paid tutors who worked in the program the previous year and were not eligible for FWS.) However, volunteerism was encouraged. Professors campus-wide gave extra credit for working in the program and tutoring met the service-learning component of several courses at Towson University. The Honors College promoted participation in the program as fulfillment of their 30-hour reflective service-learning requirement. The mathematics faculty strongly encouraged students in teacher education to become tutors, and participation was one of several service-learning options required for mathematics methods courses.

Prospective tutors filled out a half-page application in the Mathematics Department that listed personal demographics and past mathematics courses. The secretary in the Mathematics Department checked tutors' transcripts to verify passing grades in math classes taken in high school and college. We only required basic math (algebra and geometry) and a minimum score of 500 on the SAT for tutor qualification. Higher math skills were seldom needed for our target population because the objective of the program was to help children who had not passed the Maryland Functional Test. Thus, our tutors predominately re-taught pre-seventh grade skills required by that test (like multiplication, division, fractions, decimals, and percents).

Before beginning work with children, tutors signed up for an 8-hour training course that taught them to respond to mathematical difficulties common to students of this age and gave them strategies useful in enriching the students' understanding of mathematics. This ini-

Figure 4.5.4. Tutors and children enjoy one of the math games that were generously donated by an outside funding source to enhance the *America Counts* program.

tial training took place early each semester on a Saturday. Tutors could not work until they were trained because so many essential facets of the program were discussed during the sessions. Some of the training materials we used were available to facilitators by going to the Math Partners website [2], but most of the information discussed on training day evolved along with the program. Training sessions included the following components:

1. Introduction to the program (discussion of the USDE initiative, the Maryland Functional Test, schools in the program, service-learning benefits).

2. "Commitment Contract," to be signed by the tutors, stressing the important responsibility and sensitive issues surrounding the mentoring of school children.

3. Math Partners [2] activities (i.e., a discussion about the hallmarks of effective tutoring and a role-playing game to re-familiarize tutors with typical math they will encounter).

4. Introduction of lab supervisors from each site, commitment to explicit work schedules, and arrangements for car pools with tutors going at the same time to the same site.

5. Introduction to "tutor mentors" from each site (see description of these leaders below).

6. Instruction in using the tutoring materials like games, manipulatives, books, software, music, and calculators. (Four groups are formed that rotate through four half-hour hands-on workshops).

7. Necessary paperwork for getting paid. (The Career Center sends a representative to explain payroll procedures and complete necessary forms.)

8. Enrollment in the TU *America Counts* website [4] and the Math Partners website [2].

9. Orders for the *America Counts* T-shirts to wear at the school when working. (Baltimore City has waived the fingerprint requirement for our tutors as long as they wear their T-shirt as identification and sign in and out at the school office. Each tutor receives one shirt free and may purchase extras if desired.)

10. Introduction to the "Points Rack," the incentive program tutors use for rewarding children with prizes like chips, candy, toys, and school supplies.

Experienced tutors were crucial to the 8-hour training and to the subsequent success of a large off-campus program like *America Counts*. After tutors have successfully completed one year of tutoring, they may be selected to become "tutor mentors." At least one tutor mentor worked at each site and assisted the lab supervisor and me. They earned an extra $0.25 per hour for this promotion, or $10.25. Their duties included the following:

1. Supervising new tutors.

2. Matching students at school sites with tutors in appropriate time slots.
3. Carrying out various administrative responsibilities (such as contacting prospective and current tutors to ensure that they attend all training activities, carrying materials and photocopies between the schools and Towson).
4. Assisting with advertising (i.e., attending the TU Part-time Job Fair).
5. Substituting for tutors who must miss a session.
6. Helping to run workshops, T-shirt sign-up and other activities at the training.
7. Helping to plan Parent Nights.
8. Collecting input for workshop topics.
9. Serving as liaisons between the tutors and myself.
10. Collecting year-end data on the tutors and students for analysis.

While carrying out their mentoring duties, these tutors continued with their regular tutoring sessions with the children. Each tutor mentor spent approximately 75% of his or her time in actual tutoring activities and the remaining 25% in additional responsibilities.

On-going Program Support and Evaluation

Aside from the devoted presence of tutor mentors at the sites, tutors were given continuing support and training throughout their tutoring experience by means of two workshops held at the schools. I collaborated with the tutor mentors to address specific challenges the tutors were encountering in working with the students at their respective schools. Thus the real needs of the tutors and the children were able to be addressed at these sessions and the topic content varied. The topics from the last two years included hands-on methods to teach fractions, ways to reduce math anxiety, and using manipulatives and games to enhance lessons.

In addition, we maintained an *America Counts* web-based discussion site [4], which was monitored by TU faculty members within the Department of Mathematics. This Web-based discussion site was a resource for both tutors and children. Tutors posted questions about real-life mathematics applications to use with the students or asked about tutoring strategies. Children at the *America Counts* schools answered the "Problem of the Week," giving them additional contact with the tutors and another forum for learning mathematical concepts. Use of the website thus served to familiarize impoverished school children with technology and web-based resources. The website was hosted by Blackboard, Inc. at a minimal cost to the *America Counts* program.

So that teachers at the *America Counts* schools were able to respond to their students' needs during the regular school day and at other times outside of the tutoring sessions, I provided two in-service workshops at the school sites during the year. These workshops were in conjunction with the tutor workshops mentioned previously. As a mathematics education professor, I developed the materials used at the teacher workshops, and various Towson University mathematics faculty members made the presentations.

Each year a Parent Night was held at each of the school sites. At these events, I answered parent questions, presented strategies for enriching children's understanding of and enthusiasm for mathematics, and provided updates on children's progress. The parents in attendance were given a book containing mathematical problems they could do with their children at home. This book, "Helping Your Child Learn Math at Home," was especially designed and printed for distribution on Parent Night and was free of charge (compliments of TU Marketing Department.)

At the end of the school year, all tutors, tutor mentors, and lab supervisors joined together at a picnic. This dinner was designed not only to recognize and thank the tutors and tutor mentors for their work with the students, but also assisted me in evaluating the effectiveness of the program. Tutors were asked to share their experiences by completing an anonymous qualitative evaluation form. The picnic also served as a tool for retaining tutors into the next school year. Several community businesses donated prizes like manicures, oil changes, clothing, movie tickets, and the like, which were awarded at this event. In addition, all tutors (including those earning service-learning credit) were required to post an end-of-year reflection on the TU *America Counts* website [4].

Results and Accomplishments

From its modest beginning, *America Counts* at Towson University continued to grow (see Table 4.5.1). We had an excellent retention rate, and tutors seldom left the program until they graduated from the university. Most volunteers remained active in spite of having fulfilled previous service-learning quotas or pre-service teacher practicum requirements. Word-of-mouth recommendations from current tutors accounted for a large percentage of new growth. The end-of-the-year evaluation forms, reflective journals, and website reflections provided the explanations for these recommendations.

Tutors endorsed working with *America Counts* for several reasons. In the paid job category, tutors reported various benefits. First, we paid a good wage, especially in a Federal Work-Study market that is comprised solely of full-time college students working part-time. Second, the tutor was actually paid for five hours when they were physically only at the school for three hours. As mentioned previously, tutors got one hour of transportation and one hour of teacher planning time on their timesheets for every three hours they tutored. In spite of the fact that they actually drove up to an hour to and from the schools and spent the hour planning effective tutoring sessions (albeit at their home computer reading the Math Partners materials), the tutors considered these extra hours a definite windfall.

The next category of reasons that students cited as beneficial came from both the paid tutors and the volunteers who were earning service-learning credit. They mentioned the

Table 4.5.1. Number of People Working in *America Counts*. Participation in the *America Counts* program gradually increased each year. In particular, students became more involved as their friends told them about their experiences tutoring children in nearby Baltimore.

Year	Federal Work-Study Students	Grant Employed Students	Student Volunteers	Non-Student Volunteers/ Workers	Total
2000–2001	21	0	22	10	53
2001–2002	54	29	16	8	107
2002–2003	91	20	34	3	148

compatibility of *America Counts* with their university class schedules. The seven school sites had self-selected different hours for operation for their tutoring labs. Some schools ran their lab only during the day with children being pulled from class for tutoring. Others had an after-school program that stayed open until early evening. Still others had a combination of school hours and after-school hours. Thus, our Towson University tutors easily found a three-hour slot at one of the sites that left time for travel and still fit around their class schedule. We allowed tutors to switch their work hours and site location between semesters to match their new class schedule, although the occurrence of switching was rare. Instead, tutors tended to prefer their familiar carpools and work schedule and consequently registered for classes in the next semester so as to maintain the status quo.

Finally, tutors valued the *America Counts* program because they perceived this opportunity as "something where I can make a difference," whether the job was paid employment or a service-learning commitment. The following comments were made on the written evaluations from the picnic or posted on the website:

> I am glad that I was able to be a part of the *America Counts* tutoring program. Because I am going to be a teacher someday, I found this experience very valuable. I feel that the tutoring we provided definitely helps the students and is something that needs to be done in order for those students to pass the MFT.

> I liked the whole experience, and having hands-on experience with the students was really great. Some of the kids were really smart and just needed a friend to talk to.

> Thank you for giving me the opportunity to be involved in the *America Counts* program. I've enjoyed every minute of it!

> Thanks to the tutor mentors that were already there and helped me adjust always answering or trying to help me find a way to help the students. I can't believe how much I learned about math and how to apply it.

> My experience is something that I will not forget, especially the reactions from the students when they finally understood something. It was truly an experience that was very different from any school where I worked before. Thanks for making my wanting to be a teacher mean even more, there are so many children who need help!!!

> Tutoring has been such a blessing to me! THANK YOU so much for the opportunity.

> I learned so much about myself, about children, and about mathematics from tutoring.

Pre-service teachers in particular seemed to learn from the tutoring experience. Excerpts from their reflective journals in methods classes showed an increased awareness of the variety of ways that children approach mathematical topics. Many wrote that tutoring enhanced their ability to tie the theoretical concepts learned at the university to the practical application of teaching. Over 50% of Elementary Education majors reported that their *America Counts* experience had a more significant impact on their ability and confidence to teach mathematics than any other experience provided by the university. Mathematics majors echoed this sentiment in their reflections on questionnaires.

Aside from the benefit to Towson University students in the way of financial support, personal satisfaction, and increased learning, as highlighted by the statements above, the *America Counts* program has benefited several other academic arenas at the university. In

May 2003, a graduate student presented her Master's Thesis in Applied Mathematics [5] by analyzing the data collected during the program. In 2004 students in a 400-level undergraduate course plan to use data from the project as fodder for their own applied mathematics project. Real-life analysis prepares these students for applied mathematics and is more motivating as course material than the "hypothetical" situations in many textbooks.

The data analysis yielded some interesting results regarding the effectiveness of the program. Although participating schools were among those with the lowest Maryland Functional Test scores in Baltimore, significant improvement occurred at all schools. The average improvement was 14 points in seven months (a passing score being 340). On average, 31% of students in the *America Counts* passed the MFT in 2001–2. This compares to a pass rate of 16% in the control group that received no tutoring. Sixth graders showed the highest improvement rate. The percentage of eighth graders passing the MFT is of particular interest because of the requirement for high school admission — thus the pass-rate of 32% for this shrinking pool of test takers was significant.

Using a regression plot, we found that the day of the week children received tutoring was not a factor in their performance. This information was contrary to teachers' opinion that students were generally inattentive on Fridays. Another interesting result was that girls and boys were not significantly different in pass rates. A more sophisticated analysis indicated that school and grade level might be factors in performance on the MFT.

Beyond the benefits to our university tutors and the middle school students they serve, the benefits to the university itself were significant. The growth of the *America Counts* program increased the funding made available to Towson University Financial Aid Department from the Federal government. The increased funding enabled us to help more students meet the financial requirements of attending college. TU began operating its *America Counts* program solely with FWS funds in the fall of 2000. We accounted for 12% of the FWS budget, surpassing the USDE's 7% requirement. In its second year of operation, *America Counts* paid its tutors salaries accounting for 21.4% of the University's $500,000 total FWS budget. Thus in 2002 the university's FWS award was increased to $1 million by the Federal government due to this commitment to community service placements — an incentive for other universities desiring to meet or exceed the requirement of USDE. In 2003, *America Counts* tutors were paid a total of 19.3% of the new $1 million FWS budget.

Another positive result for the university was the outreach aspect. Towson University gained greater name-recognition in our community. We attracted more talented athletes, more funding for research, and more political clout. As a service-learning component, our pre-service teachers gained valuable insight through this fieldwork early in their careers. Their participation led to the development of a deeper understanding of how children learn mathematics, which better equipped them for their future vocation. Volunteers who were not education majors learned about children, schools, and the personal growth that comes from helping others, which gave them opportunity to become better parents and citizens.

Difficulties

America Counts and similar programs will always have their challenges. One of the most frustrating for us was the fact that we were working in an urban environment where absenteeism is high, and so is teacher disillusionment. We learned to make a back-up list of children who needed tutoring and then a second back-up list to ensure that tutors would actually have a student to teach while they were at the schools. In addition, teacher attendance at workshops is sporadic at best, even with the incentives of take-home kits, refreshments, and a cash stipend. The lab supervisors can be negligent in sending timesheets to payroll in a timely manner, resulting in delayed paychecks for our tutors.

The tutors also had issues — mostly related to their age and to their not-yet fully developed sense of responsibility. One of the most common problems was carpool failure and the resultant absenteeism. If the driver of the carpool decided he or she could not work (usually due to the need to prepare for a pending paper or test), all of his or her riders were stranded at Towson with no means of getting to work. This caused disappointment for the school children who waited for no-show tutors, as well as frustration for the paid tutors who did not have sufficient money in their next paycheck because they did not work. A less frequent problem was also related to irresponsibility: tutors who came unprepared to teach. If tutors had not studied the Math Partners materials or planned in advance, tutoring became a mere chat session and a waste of precious academic time.

The 8-hour Saturday training sessions could also pose a difficulty. No matter what date was selected, it seemed impossible to get all the new tutors to the first training in late September (and waiting longer was a drawback since Baltimore schools begin in August). There were always some who could not possibly make that date, and we were compelled to do a second training in October. The Spring Semester training fared better, and one session in February was usually adequate. However, in one year we needed five training sessions to accommodate all the tutors. These multiple trainings would not be an issue if the trainers were paid, but for those who were volunteering their time, these repeated sessions created friction.

A final challenge is the fact that a program like *America Counts* takes a lot of energy to coordinate. Although grants offer minor financial compensation to selected partners, like teacher in-service workshop leaders, at some point people are compelled to invest their time. In particular, the person in charge—for there must be an individual who assumes responsibility for directing the program—must either be paid or be willing to donate large quantities of time. Hiring a FWS student or a graduate assistant would ease the frequently burdensome administrative duties of the program.

One might also anticipate future difficulties due to the fact that *America Counts* dates from the Clinton administration. Future administrations may or may not be interested in the program. For example, the Bush administration has not supported these USDE programs financially except to talk of raising related FWS community service quotas. The *America Counts* [1] and Math Partners [2] websites remain active, although they have not been updated recently.

Reflection

Why would any university or individual want to get involved in a complicated combined service-learning/paid employment project like *America Counts*? Some of the concrete benefits to the university, the department, our students, and the community have already been described. But for me, the greatest reasons are those which are less tangible yet more satisfying. As a professional educator, I am in the business of building better people, better citizens, and a better society. Service to others and reflection on the lessons learned from that service are undeniable stimuli in making less selfish, more compassionate human beings—a commodity this world desperately lacks in the hustle and bustle of the twenty-first century. Healthy changes occur in the behavior of university students and urban youth as they learn about each other's worlds, and that is all the motivation some of us need.

References

www.ed.gov/americacounts for information on the *America Counts* initiative, including organization, curriculum, etc.

www2.edc.org/mathpartners for free tutoring materials available to registered participants in the *America Counts* program

Maryland Higher Education Commission (MHEC) offers Gear Up grants as part of the College Preparation Intervention Program, Annapolis, MD

Towson University's *America Counts* website is located at coursesites.blackboard.com/ as Math 205

Blum, S. (2003). *America Counts* program, Analysis of Data School Year 2001–2002, Towson, MD: Towson University.

About the Author

Candice L. Ridlon is an Assistant Professor of Teacher Education at Brigham Young University in Provo, Utah. She previously taught Mathematics and Mathematics Education courses at Towson University in Towson, Maryland. Her interest in children and learning is a result of her career as a mathematician, a certified teacher, and a mother of six children.

4.6

Can We Teach Social Responsibility? Finite Mathematics Students at an Urban Campus Tutor At-Risk Youth

Richard A. Zang
University of New Hampshire

Abstract. Over a five-year period, college students enrolled in finite mathematics at an urban campus were given the option to tutor inner-city at-risk youth. The college students sought to inculcate in these at-risk youths an appreciation for the difference between merely having facility with symbol manipulation versus actually understanding why various rules apply to a given problem; in so doing, they themselves were taught the lesson. Moreover, their ideas of social responsibility were influenced by contact with the at-risk youths they tutored.

Background

The service-learning project reported here is based at an urban commuter branch campus of the state university. The campus is located in Manchester, NH, the state's largest city. The course context is that of finite mathematics, the mainstay of the university's general education requirement in quantitative reasoning for liberal arts majors. All too often the students in this course lack many of the basic skills necessary to study the three main topics—probability, statistics, and linear algebra—although they usually have grades from high school which suggest otherwise. Indeed, research into the shortcomings of traditional testing [3,4] underscores the point that students often receive high grades absent a clear understanding of even the most fundamental aspects of the material. The university has been reticent to implement its own placement test, and in any event, a passing score on such a test would still not necessarily indicate an understanding of the material.

In addition, the students also generally lack a strong motivation to improve their basic skills. At one point, a statistical research project was introduced as an option in the finite mathematics course. Since students could choose a topic related to their discipline, this succeeded to a certain extent in addressing the motivational issue, but it did not cure their lack of basic skills. Consequently, the search began for a way to address both issues: inadequate preparation to study finite mathematics and a lack of motivation to improve their grasp of the fundamentals.

A Service-learning Option Is Born

Recognizing that a service learning environment can sometimes help students discover new points of view about both subject matter and their own abilities, I gave my students from finite mathematics the option to tutor General Educational Development (GED)-level mathematics to at-risk youth from the local community. Participating students tutored a minimum of 25 hours during the semester. They also were required to maintain a journal. The option was worth 25% of their course grade.

Operating under the belief that we cannot teach social responsibility without giving students the occasion to exercise the same, I saw an opportunity to address two important social aspects: (1) that diversity can challenge preconceived misconceptions, often based on stereotypes; and (2) that we learn from those whose experiences are different from our own. Indeed, Goodlad [2] speaks directly to this philosophy:

> Tutoring gives those who act as tutors the opportunity to learn how to care for other people. Moral education can all too easily become teaching *about* responsibility without the learner ever having the opportunity to exercise any responsibility. By involving students and children in the teaching process, tutoring can provide a valuable occasion for Study-Service which can later be developed into other activities related to the disciplines of various school or college subjects. (pp. 18–19)

Accordingly, I named this tutoring opportunity the *Social Responsibility Option*.

In order to encourage serious reflection about their activities, I required the students to maintain a journal, as is the case in most service-learning projects. The students were instructed to make entries in a contemporaneous fashion, including how they were feeling about the tutoring experience; how their thinking or feelings were changing as a result of the experience; any math-related learning encounters or insights; and how, if at all, this experience was affecting them as students in finite mathematics.

Grading Issues

Grading the *Social Responsibility Option* posed specific challenges, especially in considering how to weigh the option in relation to the other areas of evaluation within the course, and how to assess the students' individual experiences. In the early days of offering the *Social Responsibility Option*, if students did not choose it, their final exam was worth 45% of their course grade; alternatively, if students did choose it, their final exam was worth 20%, the *Social Responsibility Option* being worth the other 25%. This proved to be a poor idea; it became apparent that some students chose the *Social Responsibility Option* not because they had altruistic tendencies guiding their decision to volunteer, but rather, because they did not want their final exam to be so heavily weighted. For such students, minimal commitment translated into minimal engagement—and minimal engagement translated into minimal learning. Not surprisingly, several of these students failed the course.

Table 4.6.1. Course grading structure with and without the *Social Responsibility Option*.

Percentage totals with *Social responsibility Option*		Percentage totals without *Social responsibility Option*	
Social responsibility Option	20%		
Quizzes	20%	Quizzes	25%
Exam 1	20%	Exam 1	25%
Exam 2	20%	Exam 2	25%
Final exam	20%	Final exam	25%

The grading schema shown in Table 4.6.1 proved, by far, to be the most successful format. Although this grading model did not entirely eliminate the dilemma of students choosing the *Social Responsibility Option* for the wrong reasons, it certainly mitigated the problem. I readily acknowledged that the grading for the *Social Responsibility Option* was necessarily subjective, which I reinforced in the course syllabus, making no pretense that it would be objective. Because it would be necessary to differentiate between different levels of commitment and participation, students were reminded that their journal would supply the primary window for viewing their experience. Therefore they needed to be sure to make detailed contemporaneous entries right after each experience rather than wait until near the end of the course and complete the journal as though it were a term paper with a single due date.

The Community Partnerships

Over the five-year period during which the *Social Responsibility Option* was offered, four partnerships were formed with local community organizations. Three such associations utilized after-school tutoring programs: (1) The Alliance for the Progress of Hispanic Americans; (2) The Salvation Army; and (3) The Manchester Boys and Girls Club. The fourth partnership was with Manchester YouthBuild Odyssey, an alternative high school program designed to help former high school dropouts learn a trade and obtain a GED. Many of the at-risk youths at Manchester YouthBuild Odyssey had experienced problems with drugs and, as a consequence, the law; as one of my students would later write in his journal, many of these youths are quite familiar with the inside of a courtroom.

The culture differed greatly among the organizations and, at times, from one semester to the next, even *within* organizations. For example, over one summer, Manchester YouthBuild Odyssey got a new director, new teachers, and a new staff. Even their facility had changed (to a halfway house) and relocated to an entirely different part of the city. More importantly, Manchester YouthBuild Odyssey had developed a totally new philosophy guiding their program. The result was the functional equivalent of a completely new partnership. One substantive change was something Manchester YouthBuild Odyssey called *Mental Toughness* (part of their curriculum designed to address general study and motivational issues), which now would run during the first three weeks we had heretofore scheduled for tutoring. I had planned everything that fall semester around partnering with Manchester YouthBuild Odyssey, but I had not contemplated this relatively drastic turn of events that had occurred over the summer. I quickly learned that small grant-funded operations can suddenly fold or drastically change form.

Consequently, I had to either adapt or cancel the option that semester. I adapted by creating a role for my students in *Mental Toughness*, designing and running workshops. One example was a city-wide scavenger hunt called *The Challenge*, in which Manchester YouthBuild Odyssey students solved, among other things, mathematical conundrums created by my students. Another was a workshop entitled, *Why Smart People Fail*, wherein my students shared some of the trials and tribulations from their high school days and offered thoughtful techniques for optimal success.

Frequent unscheduled visits at the community organizations became part of my routine. On such visits, it was not uncommon to find my students helping the at-risk youths with their English homework, a science project, or even a social studies assignment. Later, in private, I would remind them that they were receiving college credit in a *mathematics* class, and to that end, why they needed to restrict their tutoring to mathematics. To avoid mathematics, it was not uncommon for the at-risk youths to attempt to deceive my students by telling such stories as, "We don't have any math homework today," or, "We already did our math homework."

To avoid such manipulation, I learned to train my students with certain prompts; e.g., if a tutee says, "I'm all done with my math homework," then the tutor might say, "Good, then let's review what we did last week." Fibs notwithstanding, I eventually learned to be more flexible—allowing them to tutor <u>science</u>—because there were periods, with *legitimate* reasons, that my tutors would not have been able to stay busy, and yet, the tutees still needed help with their studies. I reasoned that allowing them to tutor science, as well as mathematics, was in keeping with the spirit that originally motivated the *Social Responsibility Option*. This notion is supported by Brown [1]: "[I]n order to broaden the nature of mathematical experience for the student, we need to loosen the categories conventionally associated with mathematics, if we are to enable to [sic] the student to develop intuition and other ideas not easily captured in verbal or written description" (p. 64).

Other Programmatic Aspects

Three years after my program's inception, I was honored with the first-ever Dean's Award for service to the community and the college. Subsequently, several other professors (all from liberal arts disciplines) began to offer service-learning options in their courses. Unfortunately, because of lack of internal institutional coordination, some of these faculty developed relationships with the same community organizations, but they set widely different expectations for their students. For me, as a mathematics instructor, it was vital that the students do their tutoring in mathematics or in some closely related field that used mathematics. However, one can easily understand that in other disciplines (e.g., several social sciences), where the service-learning experience might be relevant in a more diffuse fashion, it might not be necessary to be so restrictive with respect to the students' activities. This led to some friction and misunderstanding with my students, who felt that I was being overly restrictive in trying to control the cope of their activities at the partner organization. This highlights the need for internal coordination and for clear communications as well with the community partner.

As time went on, I saw new opportunities for service-learning, and I incorporated them into the *Social Responsibility Option*. Some semesters, due to the nature of the organization, and/or the number of students who volunteered for the option, my students tended to pair up with certain tutees on a regular basis. During such semesters, my students took responsibility for their own learning and that of a specific tutee. I required that they each design two mathematics-based learning experiences for one tutee based on a vocational or academic component of that tutee's program. This entailed designing an alternate way to teach a given topic. Additionally, the students were required to submit, in writing, what they planned to teach, what material they needed, and what they expected in terms of measurable performance. After completing the activity, they submitted a reflection on the experience (separate from their normal journal entries), discussing whether or not they were successful, how they would modify it, and what they personally had learned as a result of it.

In the fall of 1998, while partnering with Manchester YouthBuild Odyssey, three tutors volunteered to be the subjects for an in-depth study that included three scripted interviews and three essays written in response to the question, "What does it mean to be socially responsible," and the completion of entry/exit mathematics attitude surveys. The attitude survey consisted of 59 questions which were rated on a 5-point Likert scale. With regard to the impact of the *Social Responsibility Option* on my students, the study sought to illuminate their: (1) attitudes toward mathematics; (2) views of themselves as learners of mathematics; and (3) changing definitions of social responsibility. Since students received college mathematics credit for their participation in the *Social Responsibility Option*, an important goal of this service-learning endeavor was that they develop as learners of mathematics. In particular, while the mathematics they tutored would be considered basic to them, this study sought to document changes in their appreciation for what is involved in teaching mathematics and in their perceptions of their own competence in mathematics.

Students' reports of their experiences were diverse. Their understandings of what it means to be socially responsible, their sense of mathematics, and their sense of themselves as learners of mathematics grew in ways which reflected both their individual backgrounds and their approaches to the *Social Responsibility Option*. The mathematics attitude survey data indicated that one student's confidence in mathematics increased significantly. She also enjoyed a corresponding increase in her mastery of college mathematics and found her math anxiety greatly mitigated. The other two students had felt confident in their mathematical ability at the start of the semester. Nevertheless, they all quickly realized that, although they were themselves competent in arithmetic, facility with the mechanics of solving elementary problems did not translate into an ability to teach or to explain why something works. See [5] for more detailed results and discussion.

Conclusion

The aggregated experiences of my finite mathematics students over a 5-year period indicate that, if students volunteered to tutor for the right reasons, the *Social Responsibility Option* did address the issues which led to its creation. Those students cultivated a better understanding of the basic skills needed to study finite mathematics, and developed motivation to deepen that understanding. Through seeking to inculcate in their tutees an appre-

ciation for the difference between merely having facility with symbol manipulation versus actually understanding why various rules, properties, and laws apply to a given situation, they themselves were taught the lesson. Having to elucidate these concepts, however elementary they may have been, forced my students to examine their own understanding of GED-level mathematics (or lack thereof).

Furthermore, students' ideas of what it means to be socially responsible were pro-
foundly influenced by their contact with the at-risk youths they tutored. Through significant contact with people whose experiences were different from their own, their preconceptions were challenged and they developed an increased understanding of—and thus,
appreciation for—people from varied backgrounds. The *Social Responsibility Option*
helped students communicate more effectively, and this also served the dual purpose of
preparing them for good citizenship in an increasingly diverse society. These comments
apply equally well to the at-risk youths; i.e., by having frequent one-on-one contact with
university students, these inner-city at-risk youths were exposed to positive role models,
which is a key component in helping them prepare to be good citizens. Finally, all the
aforementioned outcomes of the *Social Responsibility Option* helped to build a stronger
sense of *community*, thus reminding my students of the true meaning of that word.

References

1. Tony Brown, "Intention and significance in the teaching and learning of mathematics," *Journal for Research in Mathematics Education,* 27 (1996) 52–66.
2. Sinclair Goodlad, *Learning by Teaching: An Introduction to Tutoring*, Community Service Volunteers, London, 1979.
3. Raymond S. Nickerson, "Understanding understanding," *American Journal of Education*, 94(2) (1985) 201–239.
4. Regina M. Panasuk and Joseph W. Spadano, "Homework review model: A means of advancing students' understanding," *New England Mathematics Journal,* 31 (1998) 45–53.
5. Richard A. Zang, Timothy Gutmann, and Dawn M. Berk, "Service learning in liberal arts math-ematics: Students tutor GED-level mathematics to at-risk youth," *PRIMUS (Problems, Resources, and Issues in Mathematics Undergraduate Studies)*, X(4) (2000) 319–335.

About the Author

Dr. Richard A. Zang is an Associate Professor of Mathematics at the University of New Hampshire.
He received his doctorate from Rutgers University and was a visiting fellow at Princeton University.
Prior to entering the academy, he spent 12 years as a research scientist at IBM's Glendale Lab
Research and Development Center and RCA's David Sarnoff Research Center.

4.7

Math Carnivals:
A Celebration of Mathematics

Jane K. Bonari and Deborah Farrer
California University of Pennsylvania

Abstract. Looking for a creative way to introduce teacher candidates to K–12 students and how they learn mathematics? Try having a math carnival. Math carnivals provide the teacher candidates with an opportunity to work with the K–12 students in small groups, thus alleviating some of the anxiety that may be present concerning pedagogical skills or the math content itself. We describe how math carnivals can be planned and some of their benefits for teacher candidates, K–12 students, and in-service teachers.

Introduction

In the spring it is natural for thoughts to turn to baseball, softball, and upcoming summer vacations. However, at local elementary schools in our area, thoughts also turn to mathematics. For the past seven years teacher candidates in the Elementary/Early Childhood Education Department have helped elementary students celebrate mathematics with hours of interesting activities. The event has grown from involving only two grades in a professional development school to involving entire schools, from being a spring event to being one that can occur at any time during the year. The carnivals are now also being used to prepare for state assessment tests and/or to celebrate after the tests have been completed. The setting varies from school playgrounds to gymnasiums to classrooms.

The first math carnival was simply that, a carnival, a day of fun. A group of teacher candidates went to an elementary school and led the children in games outside in the school yard. It has since then grown into a more meaningful performance assessment task for the teacher candidates, a motivating activity for the elementary students, and a form of staff development for the elementary teachers. The following excerpt from Van de Walle [6] captures the purpose of the math carnivals quite well. The teachers that are referred to are both pre-service and in-service.

Math Carnival: Everyone must participate in at least one math carnival. This should be done in the following steps:

1. Decide which grade you would like to teach.
2. Read the chapter of Yardsticks by Chip Wood that gives characteristics of that age child.
3. Write a two page summary of these characteristics.
4. Sign up to observe the regular classroom teacher teach a math class.
5. Observe the teacher and write an observation following the given format.
6. Use the internet, books, periodicals, etc. to help you develop an active, authentic strategy that would be grade appropriate and standards based to teach to groups of eight children at the carnival.
7. One week before the carnival hand in the lesson plan to the instructor for approval.
8. After your plan has been approved, start to collect and make any materials needed for your booth.
9. Teach your strategy to your peers on the day the concept is taught in class. Make any changes according to suggestions from peers and instructor.
10. Teach your strategy to six groups of eight children each on the day of the carnival.
11. Write a reflection of your experience at the carnival.
12. Write a summary reflection of all of your experiences (yardsticks, observation, carnival).
13. Turn in the completed assignment within one week after the carnival.

Figure 4.7.1. Assignment from course syllabus.

"Every child can come to believe in this simple truth [math makes sense] and should leave school confident in his or her ability to understand and to do mathematics. The same is true for you as the teachers of these children. When you believe that math makes sense and that you are capable of making sense of it, then you will be in the best position to help your students do mathematics with understanding and with confidence."

As suggested in Figure 4.7.1, the math carnival assignment has become a way of determining if the teacher candidate has developed an understanding of the children, the math content, and the pedagogy necessary to teach that content.

The coming together of teacher candidates and elementary students enables both groups to develop positive attitudes toward their own abilities in mathematics and toward math in general. Teacher candidates have the opportunity to field test the activities they have developed by teaching them to small groups of children at the carnival. All of this happens in an atmosphere of fun and excitement, and cooperation between students in the groups helps to enhance learning. The math carnival demonstrates how university faculty, teacher candidates, in-service teachers and elementary students can learn more both cognitively and affectively through cooperation.

Objectives

It is important to keep in mind the multiple objectives of these programs for the various groups involved, namely, the university students, the elementary school students, and the elementary school teachers.

The university students are given the opportunity to teach small groups of elementary students (some for the first time) in a very nonthreatening situation. Through the use of journals, books, and websites, they endeavor to find and adapt strategies that are fun, active, and appropriate for various grade levels as dictated by state and national math standards. This practice teaching will help teacher candidates to understand the content and method for each concept they will teach and to gain confidence in their own math ability. In addition, another important objective is for the teacher candidates and university faculty alike to recognize the value of sharing their math skills and resources with the community.

The objective for the elementary students is twofold: to develop a positive attitude about mathematics and to discover that math is a part of everyday life. Figure 4.7.2 contains a sample lesson plan from a fifth grade math carnival. This plan combines geometry concepts and measurement concepts appropriate for 5th grade. It also contains opportunities for both the students and the teacher candidates to learn something about aerodynamics in a simple way.

The objective for the in-service teachers is one of staff development. Teachers observe the teacher candidates as they lead a wide variety of activities that they have adapted from resource books or the internet. Many teachers have reported back that they have used some of these activities in their own classrooms in later years. To help this happen, we compile a booklet of lesson plans that were used at each carnival, and we present it to the teachers who were involved.

Shapes in Flight

Objectives: Using written directions students will construct paper airplanes.
Students will identify any triangles and quadrilaterals found in their constructions.
Students will predict, estimate and verify the distances the planes will fly.

Materials: 8 x 11 sheets of paper, copies of directions, model of an airplane, dry erase boards, scissors, poster of various angles and quadrilaterals

Procedure:
Introduction: Show students an airplane model, point out various triangles and quadrilaterals in its construction. Let students handle and observe model.

Development:
1. Have students identify different triangles and quadrilaterals.
2. Divide students into 2 groups. Hand out paper airplane directions. (Each group will have a different type of paper airplane to make.) If somebody in your group can't understand directions on handout, help them.
3. As students make folds, have them write down the kind of triangle or quadrilateral they observe. Monitor work.
4. After the planes are constructed, have students share the kinds of triangles and quadrilaterals that they found.
5. Briefly discuss how a plane gets airborne. (Low pressure over the wing-lift.)
6. Ask students, "What plane do you think will fly the farthest?" Have them explain why.
7. Have students predict the distances their planes will fly.
8. Have the students fly their planes and estimate how far they flew.
9. Have the students measure the distances that their planes flew.

PA Math Standards:
    Standard 2.9.5: Classifying triangles and quadrilaterals
    Standard 2.3.5: Estimate, refine and verify specified measurements.

Figure 4.7.2. Sample lesson plan from a 5th grade carnival. (Source: Student generated.)

Reactions of the Teacher Candidates

Considerable growth in self confidence occurs during the days of planning and carrying out the math carnival. At the beginning of the semester, teacher candidates are usually very apprehensive about managing the children, about their own math ability, and also about teaching the math content. As they do their observations, they tend to decide that they can probably handle the children, but they are usually still concerned about whether they know the content well enough to actually teach it. Then during the carnival, they tend to relax and make the discovery that they actually can do this thing called teaching largely because of the careful preparation they have done. After the carnival we hear and read a wide variety of statements varying according to the developmental level of the teacher candidates. Those who have had more teaching experiences have more insightful comments that show their own acceptance of the responsibility for the students' learning or failing to learn what they had taught. Teacher candidates with less experience tend to notice all of the outside causes that could have affected the success of their lessons and aren't yet ready to accept any of the responsibility themselves. Occasionally we have students who have worked on several math carnivals during several school terms. It is always rewarding to see how much they have grown as they have more teaching and reflecting experiences.

Planning and Challenges

As in all projects of this magnitude, there are difficulties that can occasionally arise. However, the problems that we have encountered have generally been small ones that can be overcome relatively easily. For example, time is always a problem when trying to juggle teaching assignments, other university duties, and also trying to schedule the teacher candidates going in and out of the public schools at the same time. Most of the math carnivals that occur during the school year occur in professional development schools (PDS.) These are the easiest ones to plan and carry out because we have teacher liaisons in the schools that help with the planning. When we work with a school that is not a PDS, we must do all of the coordinating of schedules for the teacher candidates and also for the elementary schools. This semester we also had two carnivals for preschools for the Early Childhood Math Methods students.

Funding can also be a challenge. Whether it is a serious problem or not depends on how the carnival is planned. We have had math carnivals that needed funds so that the children could be given prizes and snacks. Many times the PTA is able to fund snacks. We have had math carnivals that did not have prizes or snacks and actually needed no funding at all. We have been able to borrow manipulatives and measuring tools from the university or from the elementary school. We have also hired buses to transport teacher candidates when we traveled to an urban area for a more diverse experience. This, of course, was the most expensive venture. This funding came from the university itself or from a PDS grant.

Advance planning and cooperation among all participants can help the project run with ease. Planning for all math carnivals the semester ahead allows the project to be included in the syllabus. This gives the teacher candidates the chance to know exactly what is expected and helps them to work it into their schedules well in advance. Planning several

Group	Responsibilities
University faculty	1. Acquaint teacher candidates with characteristics and math abilities of students at different grade levels and state and national standards. 2. Suggest ways of finding or developing math activities. 3. Allow teacher candidates to practice their activities in the college classroom. Give suggestions. 4. Coordinate teacher candidates' schedules with public school classroom schedules for observations. 5. Put final okay on choice of math carnival activities. 6. Help teacher candidates adapt activities for diverse students who will be visiting their booths. 7. Assess teacher candidates' work with small groups of elementary students at the carnival. 8. Lead focus group carnival assessment in the university classroom and also with the classroom teachers after the carnival.
Teacher candidates	1. Acquaint themselves with the characteristics and math abilities of students at a particular grade level. 2. Acquaint themselves with the state and national standards for concept involved. 3. Observe and reflect on math lessons taught at the grade levels they will teach at the carnival. 4. Find and adapt or develop activities or games for particular elementary grade levels. 5. Practice the activities in the university classroom by teaching to peers. 6. Adapt the activities according to suggestions from peers and university professor. 7. Prepare and/or order from the elementary school or university all materials needed for the activity. 8. Arrive at the elementary school on time prepared to present activities to six small groups of elementary children. 9. Pick up any materials that have been ordered from the elementary school. 10. Adapt activities for each group that visits the booth. 11. Return all materials and clean up area after the carnival. 12. Write a reflection about the math carnival and an overall reflection about the whole assignment. 13. Participate in focus group discussion concerning outcomes of the carnival.
Classroom teachers	1. Suggest concepts for the math activities for carnival. 2. Provide opportunities for the university students to observe their math classes. 3. Explain the logistics of movement at the carnival to elementary children. 4. Give teacher candidates suggestions for improvement during and after the carnival.
PTA	1. Make name tags and booth markers and set up booths. 2. Collect and sign out ordered carnival materials from teachers. 3. Purchase/prepare snack and man snack booth. 4. Escort children to and from restroom and nurse's station. 5. Run the time clock and sound the whistle for changing booths. 6. Return materials to classroom teachers and clean up after carnival.

Table 4.7.1. Key responsibilities for various parties.

math carnivals in a semester on different days of the week and at varying times allows teacher candidates to participate without missing other classes. Having carnivals at schools that are close to the university eliminates long travel time that can also cause scheduling difficulties.

In our experience, the best time to have math carnivals is near the end of a semester, but not too close to finals. In this way the teacher candidates will have had enough time to learn the content and pedagogy necessary to successfully teach their activities.

The last part of successful planning is to be sure that all parties involved know their responsibilities. Table 4.7.1 outline the responsibilities of the university faculty, the teacher candidates, the elementary teachers, and supporting groups such as the PTA. We have also had high school students participate in the carnivals, mostly taking the place of PTA members. This has helped them experience some teaching activity, which also can help them to strengthen their own math skills.

The Reflection Process

If we look at two math carnivals we can see how two different teacher candidates might deliberate on identical math activities. In each math carnival, there is a teacher candidate who has planned an activity to reinforce one of the standards/operations requested by the classroom teacher. Throughout the lesson, the students are active and excited. Too quickly, time runs out and both teacher candidates dismiss their respective students for the next math carnival station. Teacher candidate A feels proud of the lesson, writing that the carnival turned out to be a fun event and the children were very receptive.

Teacher candidate B responds with questions in mind such as: "Why did the lesson go so well? What worked? I guess it was the use of manipulatives that made math more fun for the students. And working in groups—they always like that. I noticed that some students withdrew from the group. I wonder, were they learning by watching, or were they learning at all?"

How were the reactions of both teacher candidates similar? How do their reactions differ? Both teachers gave good math carnival activities, motivated the students, and experienced relatively few problems. Teacher candidate A was upbeat about the lesson, thought she did a good job and will probably use precisely the same approach in future lessons.

In contrast, Teacher candidate B, had deeper afterthoughts about the teaching and learning during the carnival activity. She pondered and appeared to ask herself more self-challenging questions; she was reflecting on her teaching. Further, she can use her responses to those questions to plan changes in her teaching behavior. As a result of this reflection, it is likely that her future lessons will be even better than this one.

While both of these teacher candidates are good teachers, Teacher candidate B may be better. She is a reflective practitioner; she is already developing habits of thought that will make her a wiser teacher and ensure her consistent professional growth in increasingly complex classroom settings. In the math carnival setting, as in the "traditional" classroom setting, reflection enables all of us to learn from experience [7]. It is the ongoing process of critically examining and refining teaching practice by considering the personal, educational, social, and ethical aspects of teaching and schooling [3]. Reflections enable our teacher candidates at the math carnivals to describe and think about what they do, to anticipate and solve classroom problems, and to experience continued professional and personal growth. We feel that reflection is especially important for beginning teachers, who often hold unrealistic views about the problems they will encounter; and we believe it is the university faculty's responsibility to help structure these teacher candidates' reflections. Current research findings are supportive of our beliefs (for example, see [2]).

We have found that our teacher candidates, when participating in the math carnivals and being required to reflect on the experience, have the opportunity to inquire into and analyze their own teaching behavior and the teaching methods of others. This is further reinforced through class discussions in the methods class after the carnival has taken place. The teacher candidates appear to be sincerely interested in the fine points of the art and science of teaching, and they want to learn all they can about teaching math and doing math activities with children. In this way our students begin to take responsibility for the consequences of their teaching. These deliberations on their teaching, driven in large part by the reflection process, may change their approach to teaching and their sense of responsibility for results.

Research on the effectiveness of writing reflections shows that teacher candidates find it enjoyable and that they benefit greatly from the immediate, non-threatening feedback [1,4]. Troyer [5] found that if teacher candidates are informed about the nature and importance of reflective thinking, they are even more reflective in analyzing classroom teaching situations. This is illustrated by the following excerpts from comments of education students who participated in recent math carnivals:

"A few of the students actually knew what the three kinds of triangles were: one even knew what a trapezoid was. However, a majority of them did not know either the kinds of triangles or what a trapezoid was. To keep their attention, I did have them repeat the names of the three triangles in unison and I called on students who appeared to be drifting. As far as the activity itself, they really appeared to enjoy folding and testing their paper airplanes."

"I feel that my station in the math carnival was a success. The students got involved in the activity very quickly and enjoyed watching their person climb the ladder as they answered questions correctly. In fact, some of the students were upset because they did not have more time to play my game. I enjoyed this math carnival tremendously and think that it is a great way to implement math in an exciting and motivating way."

"The math carnival was a great teaching experience for me. I not only had the experience of planning and making an activity that I can reuse, but I also had the opportunity to work with the children and test my ideas. I also had the opportunity to work with a child who has a disability. I had fun at the math carnival and felt very welcomed by the classroom teacher and her students."

"Fourth grade students should have their multiplication tables mastered, and this was a great game for the students to practice their skills. I did notice a huge difference between students' ability levels. Working at the 4th grade math carnival made me realize that even though students should have these basic facts mastered, they still should be given opportunities to practice their basic facts."

Hopefully, the combination of the teacher candidate preparation program, opportunities like the math carnivals, and supportive elementary teachers will help the teacher candidates to reflect on their own teaching and the teaching they observe. The teacher candidates need to develop attitudes essential to reflective thinking, and the math carnivals do certainly afford that opportunity. We hope the students will experience enhanced learning and teaching as they make reflection a careful, consistent part of every lesson.

Adaptations

Although this section has centered around math carnivals planned for early childhood or elementary classrooms and taught by teacher candidates, one should not think that math carnivals need to be limited to this area. Anyone who has ever taught must surely believe the old proverb that you learn more by teaching than in any other way. If you believe this, we challenge you to plan a math carnival at some level. It can be held at the middle school level, at the high school level, or at the college level. There are many different ways that it can be planned. For example, have secondary teacher candidates hold a carnival for a local middle school or high school. Have the secondary teacher candidates plan a carnival to teach non-math majors. They can develop strategies around concepts that are commonly considered difficult areas. Challenge these teacher candidates to teach the concepts in a way that relates them to everyday situations. Allow math to be creative. Having a math carnival is a creative way to strengthen the math and/or teaching skills of many different groups.

References

1. Bainer, D.L. & Cantrell, D. (1998). Nine dominant reflection themes identified for teacher candidates by a content analysis of essays. *Education*, 112 (4), 571–578.
2. Brookhart, S.M., Freeman, D.J. (1999). Characteristics of entering teacher candidates. *Review of Educational Research*, 62, 26–60.
3. Han, E.P. (1999). Reflection is essential in teacher education. Childhood Education, 71, 228–230.
4. Peters, J.L. (2000). The effects of laboratory teaching experience (Microteaching and Reflective Teaching) in an introductory teacher education course on students' views of themselves as teachers and their perceptions of teaching. Doctoral dissertation, The Ohio State University, Columbus.
5. Troyer, M.B. (1998). The effects of reflective teaching and a supplemental theoretical component on teacher candidates' reflectivity in analyzing classroom teaching situations. Doctoral dissertation. The Ohio State University, Columbus.
6. Van de Walle, J.A. (2004). *Elementary and middle school mathematics: Teaching developmentally*. Boston: Pearson.
7. Wilson, S.M., Shulman, L.S., & Richert, A.E. (1997). 150 different ways of knowing: Representations of knowledge in teaching. In J. Calderhead, (Ed.), *Exploring teachers' thinking* (pp. 104–124). London: Cassell.

About the Authors

Dr. Deborah A. Farrer and Jane K. Bonari are Assistant Professors in the Elementary/Early Childhood Education Department at California University of Pennsylvania, in California, Pennsylvania. Both regularly use carnivals and festivals based around a variety of subject areas in their methods courses. Dr. Farrer's primary expertise is in reading/language arts, social studies, and field experience. Mrs. Bonari's specialty is mathematics education and field experience. We wish to thank all of our students for participating in this project and for sharing their reflective thoughts so openly and honestly.

4.8

Connecting Mathematics to Real-World Classrooms through Service-Learning

Lida Garrett McDowell
The University of Southern Mississippi

Abstract. This section describes the development of service-learning projects used in teaching mathematics content classes for education majors. Details of a Math Trail project are included along with brief summaries of more recently developed after-school tutoring and Garden Project activities.

Introduction

Service-learning can provide opportunities for students to see the relevance of mathematics in the real world. The service-learning activities described in this section include the Math Trail, tutoring in area K–8 after-school programs, and the Garden Project. They are used in mathematics courses for elementary teachers, taught by The University of Southern Mississippi mathematics faculty: College students apply mathematical concepts learned in their courses by participating in service-learning projects. Public school students in the community also apply mathematical concepts as they solve interesting problems and engage in activities that enhance their appreciation of mathematics.

Overview and Objectives of the Math Trail

The original Math Trail was developed in 1995–96 on The University of Southern Mississippi Gulf Park campus in Long Beach, Mississippi. During the fall semester, 1996, The University of Southern Mississippi Department of Mathematics, with the support of the College of Science and Technology, developed a Math Trail consisting of 16 stations on the Hattiesburg campus. Over the last seven years, college students in Math for Elementary Teachers classes have developed age-appropriate questions for the 16 stations and directed elementary school students from four different area public school districts to solve their way through the Math Trail. During each fall and spring semester, approximately 75 pre-service elementary teachers guide about 150 elementary students through the Math Trail.

This project gives pre-service elementary teachers experience writing age-appropriate problems for elementary students, and provides elementary school students with valuable experiences solving meaningful real-world problems. Refer to the sample Math Trail assignment and rubric (Figures 4.8.1 and 4.8.2). The Math Trail activity is a required assignment in my Math for Elementary Teachers I (Math 210) class. My objectives are that pre-service K–8 teachers will:

- apply mathematical concepts that are a part of the course material;
- use the Mississippi Mathematics Framework 2000 to construct site-related, age-appropriate problems for fourth graders;

Assignment for Math 210 students:

1. **Problem creation:** You and a partner will be assigned to a Math Trail site on the Hattiesburg campus. (See attached map of campus to see where the sites are.) Each pair of Math 210 students should develop a mathematical problem that is related to the assigned Math Trail site. Problems can have more than one part. You have the option of working as a team to develop the problem, or you can work independently from your site partner for problem creation (that is, each of you can turn in a problem). The problems you develop should be *appropriate for fourth graders.* You are required to use FACTUAL information regarding the site or relate your problem to the use of your assigned site. For example, if assigned site 5 (Mississippi Hall-Science Museum), you should include information relating to artifacts found in the Science Museum.

2. **Problem design:** To assure that your problem is appropriate, refer to the fourth-grade level of *Mississippi Mathematics Framework 2000* and determine the competency and objective addressed by your problem. Include the competency and objective with your problem. The *Mississippi Mathematics Framework 2000* can be found in Cook Library on reserve under McDowell or online at the following URL: http://www.mde.k12.ms.us/acad/id/curriculum/Math/Mathframe.htm.

3. **Solutions to problem:** EACH student should solve the problem developed. Make sure you look back at your answer to make sure the answers are reasonable.

4. **Probing questions:** In addition, each student should write two probing questions to be used in guiding the students to successfully complete your problem.

5. **Solving problems created by other teams:** When final problems have been selected for each of the sites, you will be expected to solve all 16 problems prior to the date of the Math Trail.

6 **Day of the Math Trail event:** On the day of the Math Trail, you will be expected to report to your assigned place at the designated time and carry out your assigned task in a professional manner.

7. **Reflection:** You will be asked to write a reflection regarding the Math Trail experience. You should type your reflection and respond to assigned prompts.

Figure 4.8.1. Math Trail Assignment.

Name: _____ Site location number _____

You may earn up to 50 points on the Math Trail project. The points are distributed in the following way.

10 pts. Participation with fourth-grade students _____
12 pts. Reflection on Math Trail experience _____
12 pts. Creation of problem with correct solution _____
 4 pts. Creation of probing questions to be used with your problems _____
12 pts. Correct solutions for all selected Math Trail problems _____

Rubric for creation of problem (12 pts):		
12 pts	Problem has been written, which is related to the assigned site; is appropriate for fourth graders; and has complete sentences with correct grammar and spelling. Problem is creative and interesting, and is designed to actively engage students. Problem requires the students to explore, conjecture, or reason logically. Solution is correct, and fourth-grade competency and objective have been included.	____
8–11 pts	Problem has been written, which is related to the assigned site. Problem actively engages fourth graders. Problem requires the students to explore, conjecture, or reason logically. Solution to problem is correct. One or two of the following deficiencies occur: • The problem is not appropriate for fourth graders. • The fourth-grade competency and objective are not included. • Incorrect spelling and/or grammar are found in the problem.	____
0–7 pts	Problem has been written, but does NOT require students to explore, conjecture, or reason logically, OR solution is NOT correct. Up to four of the following deficiencies occur: • Problem is not related to assigned site. • The problem is not appropriate for fourth graders. • The fourth-grade competency and objective are not included. • Incorrect spelling and/or grammar are found in the problem.	____

Figure 4.8.2. Rubric for Math Trail Assessment.

- develop and use probing questions to assist fourth graders in solving problems;
- manage and motivate fourth graders as they walk across campus and solve problems on the Math Trail;

engage in written and oral reflection strengthening their ability to communicate;

develop a higher energy level for and commitment to learning.

Elementary students participating in the Math Trail can see the relevance of mathematics as they work interesting problems. Since the problems are developed using the state curriculum framework, these students have the opportunity to apply skills they learn in the classroom. If elementary students participating in the Math Trail attend college, many of them will be first-generation college students. In this project they have a chance to interact directly with college students and can thus be encouraged to focus on their education, with the goal of attending college in the future.

Problem Creation for the Math Trail Sites

As can be seen from the assignment and rubric in Figures 4.8.1 and 4.8.2, Math 210 students are assigned a Math Trail site on the Hattiesburg campus and are required to use historical details, information about a physical structure, or factual information about site use in their problems. One semester, a problem at the Science Museum site referred to the fossilized skeleton of *Basilosaurus*, an ancient snake-like creature related to present-day whales. This problem described the head of this ancient animal as four feet long and making up one-fifteenth of the total body length. The problem then asked for the total length. I have learned to encourage the college students to construct problems that engage elementary school students in observing or measuring some aspect of the campus site. For example, in a subsequent semester, the previous problem regarding *Basilosaurus* was modified to require that elementary students measure the length of the head, in order to determine the total length. Figure 4.8.3 shows elementary students at this site. My students usually learn that it is easier to hold the interest of elementary school students when they are actively engaged in collecting data or information.

Figure 4.8.3. A fourth-grade student uses a trundle wheel to measure the *Basilosaurus* skeleton at the Science Museum site of the Math Trail.

After Math 210 students create problems related to assigned sites, the instructors of the classes participating in that semester's Math Trail meet to select the final 16 problems that will make up the Math Trail, thereby changing the problems on the Math Trail each semester. One advantage of working with fellow instructors of other sections of Math 210 is having more than one student-created problem for each site. When selecting the final 16 problems on the Math Trail each semester, we consider whether the problems are interesting, and whether they demonstrate a variety of types of problems. Each site's problem is placed on laminated paper and under the Plexiglas sheet on the podium at each Math Trail campus station. Figure 4.8.4 shows a sample problem placed at the Center for International Education station.

The University of Southern Mississippi
MATH TRAIL

Site 13 - Center for International Education

Students from around the world travel long distances to attend our university.

1) Using the globe, string, and the indicated scale on the globe, determine the shortest distance Takahiro, a Japanese student, would travel from Tokyo, Japan to Hattiesburg, Mississippi. (1 inch = 660 miles). Use your string to measure the globe distance to the nearest ¼ inch.

2) Using the graph below, determine the number of students in the English class that are from EUROPE.

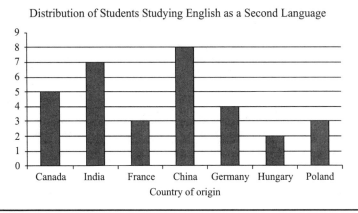

Figure 4.8.4. Sample Math Trail problem.

Logistics of Running the Math Trail

The time span for the Math Trail activities of the Math 210 students covers approximately six weeks, from the initial assignment of Math Trail sites, until the Math 210 students submit their written reflections on the Math Trail activities. Contact with the elementary school principal and participating teachers begins early in the semester in order to establish a day for the Math Trail. About three weeks before the day of the Math Trail, directions are sent to participating elementary teachers, including a request to group elementary students into teams of three to five students.

On the day of the Math Trail, elementary students arrive on campus organized in teams. College students, usually in pairs, are assigned to co-sponsor each team of elementary students, escorting them to assigned Math Trail stations. Figure 4.8.5 shows a team of elementary students and co-sponsors at M. M. Roberts Stadium. Elementary teachers are assigned to teams and also participate in the trek. Each team begins the Math Trail at a different campus site to avoid having more than one group at any one station. Pencils, paper, a calculator, measuring tools, and nametags are distributed to each fourth-grade team as they begin their trek on the Math Trail. Each team receives a sticker for successful completion of the problems at each station.

The Math Trail event itself takes about 90 minutes. Each elementary student team, along with college-student co-sponsors, is assigned to visit a subset of the 16 campus stations. A minimum of six stations is visited. We ask the school districts of the participating elementary schools to provide transportation to and from the university campus. Thus, we must schedule the activity between 8:30 A.M. and 2:00 P.M. We must also schedule the Math Trail during the college students' class times. Each fall and spring semester we typically offer three sections of Math 210. Usually, the Math Trail is scheduled during the class times of two sections, and the students in the third section are encouraged to join one of the other sections for the Math Trail activities.

Figure 4.8.5. The M. M. Roberts Football Stadium is one of 16 stations with permanent wooden podiums on the Math Trail. Math 210 students interact with fourth-grade students at this station.

Regarding funding or other support, the Department of Mathematics provides drinks and snacks for students at the end of the Math Trail. We have also worked with schools whose principals have arranged for box lunches to be sent. In these cases the Department of Mathematics has provided drinks. During spring semester, 2003, the Alpha Mu chapter of Delta Kappa Gamma provided drinks and snacks for the students participating in the Math Trail activity. The Dean of the College of Science and Technology, with the support of the chair of the Department of Mathematics, provided funding to construct the wooden podiums, which form the Math Trail stations. Each administrator was apprised of the benefits to pre-service teachers, and reminded that this activity would build good relations with area schools and prospective future students by exposing elementary students to higher education.

Reflection Component of the Math Trail Assignment

A written reflection is the final requirement of the Math Trail assignment. Math 210 students are asked to respond to three of the following four prompts.

- What did you learn from the Math Trail experience as you progressed from being a student to being a teacher of K–8 Mathematics?
- What surprised you about the fourth graders or the Math Trail activities?
- What improvements or changes need to be made in order to make the Math Trail more meaningful for the fourth graders and/or Math 210 college students?
- Describe the reaction of the fourth graders during the Math Trail.

The written reflection allows Math 210 students to formulate and communicate discoveries made during the Math Trail while providing important input regarding the effectiveness of the project.

After the Math Trail event, students are able to incorporate insights gained during the Math Trail assignment into our class discussions. Through oral reflection students have the opportunity to connect mathematics to their experiences with elementary students.

Student Reactions to the Math Trail

Several themes recur in the reactions of the college students, as shown in Table 4.8.1. Some Math 210 students are initially apprehensive about their ability to teach mathematics (Student A), but many students also gain confidence through this assignment (Student E). Many students realize, through working with elementary students in this activity, that students learn in different ways (Student B). Managing a learning activity for a small group of students is a new experience for many Math 210 students, but college students participating in this activity have a chance to practice classroom management techniques, as evidenced by the comment of Student C. The experience of working with elementary students gives Math 210 students insight into what is required of an effective teacher, as indicated by Student D.

Many Math 210 students do not readily embrace the depth with which they are being asked to approach mathematical topics in this course, as they have previously memorized algorithms with little understanding. For some Math 210 students, working with elementary students provides the motivation for the often-difficult task of learning new ways to approach mathematics, as seen from the comments of Student F.

Table 4.8.1. Sample student reactions after participation in the Math Trail.

Student A	"I was actually nervous and scared about doing this assignment. I was by myself with four little girls. I was shocked by their enthusiasm."
Student B	"I'd have to find a variety of different ways to explain a problem before they lost the blank look from their faces."
Student C	"Some kids will dominate the scene and you need to try to pick out the less dominant ones and get them to participate as well."
Student D	"Today was not only about supervising the fourth graders solving math problems; it was also about directing them and providing them with the hints to solve each problem. I felt more like a teacher than a student. Some students were able to work the problems on their own, some needed a little help and direction with each problem, and then some needed total supervision, guidance, and help with each problem."
Student E	"Somehow I was able to come up with correct probing questions to help the children with the problems. This really gave me confidence that I can succeed in teaching my students one day."
Student F	"The preparation that our class went through to prepare for the Math Trail was quite an ordeal. I know now that there is a lot of work involved in teaching students mathematics and I also need to make the information interesting to them so that they will understand and learn the mathematical topics."
Student G	"This really showed me how much we have learned in this class is being taught in the actual classroom setting. I mean there were instances that the girls used simpler problems to solve harder ones. Also the girls were using skills that we had covered such as patterns, and the problem solving techniques. This really made what we are learning a little more real since I may be teaching these concepts one day."

The Mathematics for Elementary Teachers classes are content courses in which we focus on helping students improve their conceptual understanding of K–8 Mathematics. In problem solving, we focus on developing a toolbox of strategies, using several strategies to work the same problem. In the Math Trail, Math 210 students not only realize the necessity of knowing a variety of strategies for solving problems, but they are also able to apply and observe the problem-solving strategies learned in the class (Student G).

Level of Student Accomplishment in Math Trail Activities

An estimated 75% of Math 210 students successfully create Math Trail problems that require fourth-grade students to explore, conjecture and reason logically. Pairing less motivated students with more motivated colleagues is an effective way to encourage the creation of interesting problems. Math 210 is the first Math for Elementary Teachers class taken at the university by these students, so few students are familiar with the Mississippi Mathematics Framework 2000 when they begin. Approximately 75% of the students appropriately reference competencies from this publication in creating age-appropriate problems. A majority of the Math 210 students create thoughtful, probing questions to assist fourth graders in working the problems. Some students do not understand the need for these questions until they interact with the fourth graders, but then in retrospect, wish they had created guiding questions.

By the completion of the Math Trail activities, based on student comments in written reflections, Math 210 students are generally very positive about the experience. Rarely is a student absent on the day of the Math Trail. All Math 210 students who are present participate in the Math Trail event and manage student learning through their participation. Elementary teachers whose students have taken part in the Math Trail appreciate the opportunity for their students to take part in the activities and are eager to participate when invited again. In the April 8, 2004 edition of the *Hattiesburg American*, one elementary teacher stated, "The Math Trail is always a lot of fun for our kids. It's a great opportunity for the children to apply skills they learn in the classroom."

Evolution of the Math Trail

The Math Trail has evolved as we have learned from our mistakes. In an effort to provide a service-learning component in each of the Math for Elementary Teachers classes, the college class participating in the Math Trail has been changed. In fall semester, 1996, I developed the assignment for students in the Math for Elementary Teachers II class (Math 309). Topics covered in Math 309 include fractions, ratio and proportion, percent, statistics, and probability. I chose to invite sixth graders to participate in the Math Trail because they were studying many of these same topics. College students were able to apply their knowledge of concepts studied in this class by developing questions related to these concepts. Because problem solving, numeration systems, whole number operations, and number theory are the main topics covered in Math 210, where the project is now located, material studied by fourth graders is better aligned with this course and thus these are the students who now are invited to Math Trail.

The scheduling and preparation for the Math Trail event has been streamlined as a result of input from other instructors who have collaborated in the planning of this event. Currently, all instructors teaching Math for Elementary Teachers I (Math 210) use the Math Trail activities during fall and spring semesters.

Although I began using the Math Trail before I was aware of service-learning, I have been able to make this activity a richer learning experience for students because of my involvement in seminars offered by our campus Office of Community Service-Learning. For example, I have expanded the written reflection requirement of this project, and have become more adept when interacting with students as they reflect on the service experience as a part of class discussion. The number of points earned by completion of this project has doubled, and the Math Trail has become a focal point of the class.

Recently, I used a digital camera to share photos with college and elementary students, rather than sending hard copies of photos to elementary teachers and students. Using digital photos allows me to more conveniently share photos with the elementary teachers and students and to post photos to our Math 210 class webpage.

The National Math Trail (www.nationalmathtrail.org) is a valuable on-line resource that can be incorporated into a Math Trail project and would offer an opportunity for future teachers to use technology as a tool as well. The National Math Trail is an opportunity for K–12 teachers and students to discover and share the mathematics that exists in their own environments. In particular, students explore their communities and create math problems that relate to what they find, and their teachers can then submit the problems, along with photos or videos, to the site. Students, educators and parents can access submissions, which are indexed according to grade level and math topic.

Other Service-Learning Activities: Tutoring and the Garden Project

As a result of an Office of Community Service-Learning seminar during fall semester, 2001, I developed an assignment requiring students in Math for Elementary Teachers II (Math 309) to tutor in after-school programs for K–8 students. This assignment required the college students to informally assess the mathematical level of K–8 students and develop an activity or game to meet the needs of these students. College students reference mathematical concepts used in the activity to concepts studied in Math 309, competencies and objectives in the Mississippi Mathematics Framework, and standards in the National Council of Teachers of Mathematics' *Principles and Standards for School Mathematics*. This assignment, requiring five hours of tutoring, replaced the Math Trail as the service-learning component in the Math 309 class. Since tutoring is discussed at some length elsewhere in this volume, I will not elaborate here.

The Garden Project provides opportunities for pre-service K–8 teachers to use and apply measurement and geometry topics as they interact with K–8 students. Students in Math for Elementary Teachers III (Math 310) are assigned to observe and work in the gardens at area elementary or middle schools and after-school programs for students. The Math 310 students then develop and implement measurement activities related to these gardens. These garden sites were developed and maintained through The University of Southern Mississippi's Lighthouse Partnership grant funded by the Mississippi Center for Community and Civic Engagement. I was introduced to the Lighthouse Partnership Garden Projects by the Office of Community Service-Learning and recognized an opportunity for Math 310 students to develop measurement activities related to the gardens. As in the Math Trail and tutoring experiences, college students are asked to reference our state mathematical curriculum framework and national mathematics standards. College students meet a community need by maintaining and in some cases helping to construct garden sites. The Garden Project assignment, rubric, and sample lessons developed by pre-service teachers can be found at this URL: www.usm.edu/ocsl/garden_project.html.

Objectives for After-School Tutoring and the Garden Project

Similar to objectives for the Math Trail, my objectives for after-school tutoring and the Garden Project are that students participating in these projects will:
- apply mathematical concepts that are a part of the course material;
- use the Mississippi Mathematics Framework 2000 and the National Council of Teachers of Mathematics Standards to construct age-appropriate activities;
- develop and use probing questions to guide K-8 students;
- manage and motivate K-8 students as they work in small groups;

- engage in written and oral reflection strengthening their ability to communicate;
- develop a higher energy level for and commitment to learning.

Students at elementary and middle schools and after-school programs are provided with activities that can introduce or reinforce mathematical concepts. Through these service-learning experiences K-8 students can make connections between mathematical concepts and connect mathematics to other disciplines. College students can be positive role models for these elementary and middle school students as they work with them in small groups.

After-School Tutoring and Garden Project Reflection

In order to give students a chance to communicate what they have discovered or learned in the after-school tutoring and Garden Project assignments, they are asked to write a reflection responding to the following prompts.

- Comment on the effectiveness of the lesson (activity), and discuss what could be done to improve the lesson (activity).
- Express your feelings regarding your experience in meeting the requirements of the assignment.
- Relate your experience with the after-school tutoring project or Garden Project activity to your professional development as a prospective teacher.

To give students additional opportunities for oral reflection, I allot one class at the end of the semester for students to verbally reflect on their service-learning experiences. The discussions are videotaped in order to allow for further analysis or review of student comments. Math 310 students are given the option of sharing service-learning experiences with classmates in one of the following formats:

- PowerPoint presentation including photos and student work;
- Poster presentation including photos and student work;
- Oral presentation including student work.

As college students share their learning experiences, some students thus get the side benefit of utilizing and increasing their knowledge of technology.

Student Reactions to Math 309 and Math 310 Projects

Judging from written reflections, most Math 309 students feel that five hours of tutoring per semester is a reasonable requirement. In our class discussion, the consensus among Math 310 students was that the Garden Project should be a required assignment, but no more than four hours per semester at the service site should be required.

Insights gained by individual students can spark interest and motivate them to a higher commitment to learning. As one student reflected in our class discussion, "You made darn sure that you knew what you were supposed to be doing. You made sure you knew what that was so you could tell those little kids. You didn't want to go out there and not know what you were doing." As an example of increased commitment to learning, another student wrote: "The Garden Project made me so excited to see students that really wanted to participate. The children just jumped in and went to work. Excited students make excited teachers."

Some students reaffirmed their career choices. As one student wrote, "Being around the kids, helping them with their work, and even just talking to them restated in my mind how much I want to be a teacher and that it's so right for me."

Challenges in Service-Learning Projects

The structure of university class schedules and the personal commitments of college students present scheduling challenges in each of these projects. Our college classes meet for only 50 minutes Monday, Wednesday, and Friday, which is not long enough for the 90-minute Math Trail event. In organizing co-sponsors in these classes, I am careful to place at least one Math 210 student on each team who can stay longer than the regular class period.

The Office of Community Service-Learning personnel have been helpful in setting up service sites for after-school tutoring and the Garden Project, and establishing links between university faculty and community personnel. However, community organization schedules are sometimes not convenient for college students, especially those who work, have families, or commute. In the Math 309 after-school tutoring project, we have allowed students to tutor in after-school programs in their hometowns, as well as the Hattiesburg area. Giving students ample time to schedule service-learning activities occurring outside of scheduled class time can reduce scheduling problems.

Sometimes students working at after-school programs have unrealistic expectations concerning the organization of the program. They would like to observe the same group of students, build a rapport with them, evaluate their skills and make sure their lesson plans correlate with their abilities. In after-school programs, administrators generally cannot guarantee that the same students will be present consistently when the college students are there.

Pre-service students have also been concerned when they were not prepared to help students with prerequisite concepts for the particular activity being implemented. I plan to revise project requirements to include at least a listing of skills required to complete the activity or problem so college students can prepare to teach the necessary prerequisite skills.

Projects Are Works in Progress

Currently, we include service-learning components described in this article in each of the mathematics classes for pre-service K–8 teachers. Students in these courses benefit from experiences with K–8 students early in their college careers by finding out if they really want to work with students at this level. Each of these activities can provide opportunities for students to validate their decision to teach K–8 students.

Even though one student referred to the preparation for the Math Trail, which does take several weeks, as an "ordeal" (see Table 4.8.1: Student F), the on-campus, one-time, 90-minute interaction with elementary students seems generally appropriate and rewarding for Math 210 students. After participating in the Math Trail, most students are eager to participate in the more challenging Math 309 and Math 310 projects, which require off-campus travel and 4–5 hours of tutoring/observation, along with the development of lessons or activities and their subsequent implementation.

The Math Trail has evolved over a period of seven years and is an integral component of the Math 210 course. In contrast, the after-school tutoring project and the Garden Project are in more formative stages of development. Each of these activities will continue to be modified to meet needs in the community, and to enhance student learning.

Acknowledgments

I would like to acknowledge Becky Becnel for developing the original University of Southern Mississippi Math Trail, and for sharing her materials related to this project. I would also like to acknowledge colleagues in the Department of Mathematics who have participated with enthusiasm in service-learning activities, and contributed suggestions resulting in more effective projects. Finally, I would like to acknowledge the encouragement and support of Richard Conville and Hunter Phillips in preparing this essay, and for their significant contributions to service-learning at The University of Southern Mississippi.

About the Author

Lida Garrett McDowell is an Instructor of Mathematics at The University of Southern Mississippi, where she works with students in the mathematics education program. She has also taught science at the secondary level.

4.9

Family Math Nights:
Sharing Our Passion for Mathematics

Perla Myers
University of San Diego

Abstract. Future elementary school teachers organize family math nights at local elementary schools. The result is a win-win situation for everyone involved: the parents and children experience working on interesting mathematics problems together and the future teachers realize the need for developing deep mathematics content knowledge, while reinforcing their desire to serve the community.

Introduction

A room full of children...a room full of parents...a room full of future teachers. There is a tangible feeling of community with everyone working together solving mathematics problems...and loving it! It is 6:00 pm. The elementary school auditorium/cafeteria is buzzing with sound. Tables and chairs neatly arranged... people everywhere... children working with their parents solving mathematical problems and challenges created by our children's future teachers (see Figure 4.9.1). I can't believe that 10 minutes ago the room looked like an after school day-care program! There's math talk and work all around. It's so exciting. Family Math Night is definitely here to stay!

The Challenge

When I received my teaching assignment for fall 2001, I knew that I would be embarking on a new learning adventure. The thought of teaching the course "Mathematical Concepts for Future Elementary School Teachers" for the first time made me both excited and nervous. Students taking this course are on their way to becoming our children's teachers. The course is designed to strengthen their mathematical foundations, improve their problem-solving skills, and guide them to evolve into independent learners. Yet, many students in

Figure 4.9.1. Future teachers working with children and their parents.

these courses lack positive math experiences and much basic knowledge. In fact, some of them come with years of built-up frustration and anxiety, and weak math backgrounds. Furthermore, some of these students believe that they only need to know basic arithmetic because "they will only teach kindergarten or first grade."

I had heard several stories from other mathematicians around the country about how challenging, yet important, teaching these classes can be. We need to help our students develop a facility with mathematics that will lead to an enjoyment of this important subject. After all, they will be passing their attitudes on to hundreds of children! I was also told that misconceptions about the course abound. Many students expect to learn methods for teaching children mathematics. Others anticipate a "cake" course where they will review basic arithmetic at an elementary school level. These future teachers, on the other hand, are very motivated to do anything that will benefit their future students. As future teachers, they are also eager and committed to serve their community. I was determined to change their misconceptions, if indeed they existed, as well as promote their dedication to service. I would try to provide opportunities to help them discover that it is imperative that teachers have strong mathematical concept knowledge. I would make sure that all the students in the class had some positive experiences while working hard to strengthen their mathematical foundation. Thus, the challenge: How can we best achieve these goals?

Service-Learning Involvement

Community Service Learning was definitely the answer. The Community Service Learning Office at my university encouraged me to incorporate service-learning into my classes and helped immensely in the initial brainstorming. Experts from our Community

Service Learning Office facilitated tutoring opportunities for interested students and trained our student leaders. The Community Service Learning Office provided support for the service-learning aspect of the course throughout the semester and facilitated sessions for learning, reflection, and interaction with other faculty members who were integrating service learning in their courses for the first time. In keeping with the community outreach facet of the plan, and because most of my students seemed to have a craving for seeing applications of the mathematical concepts they learn set in an elementary school context, I decided to organize a Family Math Night at a local elementary school (K–8). The students would provide a service to the community while at the same time achieving my educational goals for them.

Goals and Objectives

Each of the five Family Math events (one was held during the day) that we have held to date has been truly a "win-win-win" experience. As I embarked on the planning for one of these events, I set out some clear goals that I hoped to achieve. The goals are broken down into three distinct groups—goals for the elementary students and their families, goals for the "future teacher" students, and goals for me as the instructor.

For the elementary school students and their families, the goals are to provide positive mathematics experiences, engage the children and their parents in joint mathematical thinking that we expect will continue after leaving the event, give the children an opportunity to see that their parents value mathematics (see Figure 4.9.2), reinforce the young students' positive learning attitudes through their feeling of accomplishment at successfully solving mathematical challenges, and give the families a list of resource information, such as

Figure 4.9.2. Parents, students, and pre-service teacher working on a money problem.

books, educational software, and good websites for reference material. At the end of the event, each child also receives a mathematical book or software (all donated; see below).

At the same time, my students need to make some connections between the mathematics they learn in class and that which sparks an interest in elementary school children and their families. Making these connections requires the future teachers to appreciate the need to acquire a deeper understanding of mathematics in general. In order to accomplish this link between the real world and their classes, they have to develop and test many different activities for the youngsters and have the experience of working closely with the K–8 students. The process of developing rich mathematical activities and creating a positive math experience for the young kids goes a long way in dispelling the idea that children do not enjoy math. Rather, they get a glimpse of mathematical thinking in action, reinforcing the concepts learned in class and helping them view the subject as more than just a set of isolated algorithms. Additionally, through this event, the future teachers participate in a real-world service learning "experience", thus demonstrating the rewards of community involvement. This type of event also helps to enhance their communication skills.

I have several goals for myself as the class instructor, as well. They are mostly side-effects of the event. Mainly, I benefit by participating in the creative process involved in developing rich mathematical activities. Collectively, my students bring a wealth of ideas and great insight into the learning process of children, and we all gain when the innovative stores are tapped through our brainstorming. Additionally, students understand the need to develop a deeper understanding of mathematics and, generally, become more interested in learning and show more positive attitudes toward mathematics. As an instructor, I cherish the moments after this metamorphosis has occurred, as it results in more effective and enjoyable class time. As an extra bonus, I get to enjoy observing the wonderful interactions between the families and the future teachers and the high energy that all bring to the event (Figure 4.9.3).

Figure 4.9.3. A busy auditorium.

I will honestly say that Family Math Night met or exceeded each of these goals. It was an unqualified success! Here are some of the reasons: First, to prepare for the Family Math Night, 60% of the students tutored mathematics weekly in several elementary schools in the local area. The tutors were able to make connections between the concepts they learned in class and the mathematics encountered in elementary school. They also observed the different approaches to teaching mathematics that are currently employed, often quite divergent from their own learning experiences as young students. The students who were not able to tutor also gained from the experiences of their peers, as shared during class. The schools appreciated having the extra resources provided by the tutors and the children enjoyed working closely with college students. Here are some comments from students who tutored:

"…the concepts we are learning are the same concepts they are learning. In second grade students understand place value and number relationships as they add and subtract."

"I gained much insight into how children think and process basic mathematical concepts. I was able to witness firsthand all of the common mistakes that are mentioned in our textbook. This helped me to formulate better, more effective ways of teaching and explaining these basic concepts. I learned that it really is necessary to explain why something works before teaching the formula or "trick" that can solve a problem. It is important for me to stress how much better I understand now that children need to understand the "why" first."

"I learned so much from the kids…I could apply a lot of the ideas and concepts that I learned in class to tutoring these kids…Overall, this experience was amazing for me. By teaching others what I had learned I was helping them out but also helping myself out because by teaching them I got a deeper understanding of the ideas."

Second, over the course of four semesters, the students have used or developed more than 80 mathematical activities and presented them at the Family Math events. In order to field questions and interact with the elementary students and their families during the encounter, my students are forced to develop a stronger mathematical knowledge.

Third, and best of all, I have had at least a couple of students who "get it!" during each Family Math event. The light bulb goes on in their heads. They realize that the best way to get through to the young elementary students is with a positive attitude. After that evening, these students are motivated to succeed in my class, and hopefully, as new teachers. I am always pleased with the performance of the class as a whole, but I am particularly proud of these students.

The Procedure

Let me share my own views on how to initiate a Family Math Night, although of course others may wish to pursue variations. As with any good plan, the Family Math Night idea can be broken down into at least three distinct phases: The Build Up (what happens before the event); The Execution (controlled chaos on the night of the event); and The Aftermath (reporting on what has been done).

The Build Up

Most of the effort takes place during The Build Up phase. The most important aspect of this phase is effective communication—communication with the school, as well as communication of expectations to the future teachers.

Several months before the event...

First on the agenda is finding an appropriate place and date for the event. Elementary schools are generally receptive to involvement from the community, especially from colleges and universities in the area. Family Math Night requires little effort from the host school, and thus we have several elementary schools requesting our Math Events. We have held Family Math events in three different schools: Longfellow School, a Spanish immersion magnet; Carson Elementary, our neighborhood school, which includes a large percentage of children whose parents do not speak English (we sent flyers in Spanish, Mong, English and Vietnamese); and Hawthorne Elementary, another local neighborhood school. We generally plan the date of the event with the host school a few weeks before the semester begins so that the information can be included in the school mailings and in the course syllabus. We have found that the best time to schedule the event is approximately a month before university classes end. At that stage, the students have encountered many mathematical concepts in class that they can explore further. Yet, it is far enough away from the end-of-school rush to conclude projects, graduation, finals, etc. At the elementary schools we have worked with parent representatives, as well as directly with the principal and vice-principal. It is important to have some contacts at the school and to attend at least one parent-teacher meeting to seek volunteers to advertise and help during the event.

Before the semester begins we secure some funding and contributions to support the event. My university has been generous in awarding us internal grants to purchase materials and to provide dinner for the college students before the event and refreshments (cookies and juice) for the families after enjoying the mathematical activities. We have also requested and received donations of enough books and software with mathematical themes so that each child would take one home (some donors are KPBS Public Radio/TV, Scholastic Books, Costco, and Microsoft). The future teachers are notified of their Family Math Night commitments on the first day of class. They begin to "shop around" for a compatible working partner and for worthwhile mathematical activities. There are several resources in my office, as well as on my website, and I share activities with them during the first few minutes of class each day. On the first class meeting one or two students have the opportunity to undertake a greater responsibility for the community service learning component of the class as Student Leaders. They become an integral part of the planning and implementation process.

A month before the event...

Several weeks before the event we send a flyer to parents (see Figure 4.9.4). We ask the families to fill out and return the bottom portion of the flyers so that we can get a better idea about attendance. Each school needs and receives different types of publicity in order to encourage families to attend the event. For example at Longfellow School, where there is usually ample parental involvement, families received a flyer, an email announcement, and a reminder in the monthly school newsletter. At Carson Elementary School, the vice-principal developed flyers in the various languages spoken by the families. Each family received three flyers at different times throughout the month preceding the event. Additionally, several parents called each household, speaking in the appropriate language, and invited the parents and children. In both cases, the Family Math Nights were very well attended.

Figure 4.9.4. Flyer for the families is sent several weeks before the event.

A month before Family Math Night the students receive more details about their responsibilities (see Figure 4.9.5), and we view a video taken in previous semesters. After the first event, I have been able to give them more guidance as I describe to them successful and weak activities, and we brainstorm about ways to improve and extend them. The task of creating an activity may seem overwhelming to some students at first, but the future teachers quickly become enthusiastic as they approach their task. Each pair of students selects a grade level to target. Some pairs choose to use existing activities from teacher resources, while others develop their own activities incorporating the concepts they learned in class.

FAMILY MATH NIGHT

It's time to start thinking about our Family Math Night, scheduled for Wednesday, May 4th, 5:00–8:00 PM at Alcott Elementary School.

Our goals for the school are:
1) To provide positive mathematics experiences for the families,
2) To engage children and their parents in mathematical thinking,
3) To give children an opportunity to see that their parents value mathematics, and
4) To help students discover the fun of doing mathematics, reinforcing their positive attitudes.

Your tasks: (each pair of students)
1) Develop a worthwhile mathematical task appropriate for a specific grade (K–6th) **using some of the concepts we learned in class.** A worthwhile mathematical task:
   - is accessible to everyone at the start,
   - is extendible,
   - involves students and parents in speculating, explaining, justifying, etc
   - encourages collaboration
   - is engaging

   Families will spend about 25–30 minutes at each activity.

2) At the family math night:
   - be ready to share your activity with several families
   - provide handouts so families can remember, continue and extend the activity at home
3) After the family math night, write and turn in a report including a description of the development process of your "worthwhile mathematical task," your experience during the math family night, lessons learned, and your reflections.

Timeline:

From now until Tuesday, April 12th: Do some research on the web, math for children magazines, math books, etc, and find or DEVELOP your favorite **worthwhile mathematical task**. I have lots of resources on my website and in my office. Please feel free to come by and look and I'll be happy to help you get started. Be creative!

From now until Wednesday, April 20th: let me know what you plan to do and try out your activity with some children and some of your friends to see how it works, how long it takes, etc. Make adjustments as needed. Make sure you and your partner come by my office and get your plans/activity approved.

On Thursday, April 21st: Be ready to try your activity with fourth graders from Lemon Grove who will be visiting the university for this purpose.

From Thursday, April 21st until Tuesday, May 3rd: Work to polish your activity.

Wednesday, May 4th: FAMILY MATH NIGHT! The day we've all been waiting for! Meet at USD to carpool to Alcott Elementary School (4:45). Meet at Alcott (not too far from USD) at 5:00 pm to set up and get ready. See directions on the back.

Figure 4.9.5. Instructions for future teachers.

A week to two weeks before the event…

Future teachers discuss their ideas with me and continue to improve their activities. At this time we also send reminder announcements to the families, and we start making final arrangements including:

> *Making nametags for the future teachers*
> *Ordering dinner for the future teachers*
> *Finding out the number of children and parents planning to attend the event and securing enough space at the school*
> *Buying juice, cookies, cups, napkins, nametags for the students, and other supplies*
> *Making photocopies of handouts and materials*

Making a sign-in sheet
Confirming attendance of parent volunteers
Securing rides for the college students
Emailing reminders to the college students and to the families

A strategy that works well to entice children to attend Family Math Night is a visit to the classrooms the week before the event. Volunteers from the Math Concepts for Future Elementary School Teachers class go to the classrooms during the week preceding the event and give the children a taste of their activities.

The day of the event:

There is much to be done the day of the event. Parent and student volunteers help to set up the rooms—we have used the auditorium and cafeteria areas, as well as classrooms for the younger students. After-school programs usually occupy the space where Family Math Night occurs, so setup needs to be done in a short period of time. We set up one table for each activity, keeping the different stations for each grade (roughly five stations for the younger grades and 2–3 stations for the older grades) clustered together. Each table is given a grade label and a color. We also set up the table for refreshments, a sign-in table, and a display of the books and software to be raffled. As the future teachers arrive, they set up their areas and get ready to receive their "customers."

The Execution

All the preparation during the previous few weeks finally comes together on the night of the event when all the plans must be executed. The Execution Phase is the shortest phase, but it is definitely the most fun. Children and their parents sign in, receive a nametag and a raffle ticket, and get assigned to their first activity (this semester we used colored helium balloons and each child received a plastic lei matching the color of the balloon corresponding to his/her first table). Since the activities begin at 6:00 PM and we encourage families to arrive early to sign in, we have several tasks for the families while they wait for the booths to open. For example, this semester they had the opportunity to estimate the number of kernels in a large jar of popcorn (the closest estimate won the jar and its contents); they were encouraged to create "$1.00 words," where the value of a word is the sum of the values of each of the letters composing it (A = $0.01, B = $0.02, C = $0.03,…); and, they had the task of creating certain patterns by folding a paper and punching a single hole. Some parents became so involved in these activities that they spent the rest of the evening working on them.

At 6:00 PM, the families take their places at their assigned tables, the principal of the school welcomes everyone and the families start working together. After approximately 25 minutes the children and their parents go to a new activity. We have tried several strategies for having the families switch tables. A couple times we turned the lights off and back on and the college students directed the families to the next table. Another time we selected a student leader for each grade and she/he consulted with other activity leaders to decide on the appropriate time to switch. At Carson Elementary School, where we knew that families would arrive throughout the evening, we allowed them to flow freely through the activities and stay at each table as long as they desired. In fact, I encouraged the future teachers to try to engage the families so that they would choose to remain at their table, building on a specific concept. The college students who took part in both the free-flow-ing and the more structured evenings preferred the structured evening because they had more control over the amount of time their activity would last.

After each family takes part in three activities (approximately 80 minutes), we serve refreshments and award prizes to all the children via a raffle. As the numbers are called,

---

FAMILY MATH NIGHT MATH RESOURCES
Longfellow School

BOOKS:

Games, Puzzles and Numbers:
Moscovich, Ivan, **Number Games**
Moscovich, Ivan, *Probability Games*
Rohrer, Doug, *Thought Provokers*
Heimann, Rolf, *Mind Munchers*
Snape, Charles and Scott, Heather, *Puzzles, Mazes and Numbers*
Ryan, Steve, *Mystifying Math Puzzles*
Ferris, Julie, *Math for Martians: Planet Omicron*
Clemson, David and Wendy, *My First Maths Book*
Tyler, Jenny & Round, Graham, *Brain Puzzles*
Blum, Raymond, *Math Tricks, Puzzles and Games*
Anno, Mitsumasa, *Anno's Counting Book*
Burns, Marilyn, *Math for Smarty Pants*
Cartwright, Stephen, *First Book of Numbers*
Heinst, Marie, *My First Number Book*

Story books:
Pinczes, Elinor, *A Remainder of One*
Anno, Mitsumasa and Masaichiro, *Anno's Mysterious Multiplying Jar*
Anno, Mitsumasa , *Anno's Hat Tricks*
Anno, Mitsumasa , *Socrates and the Three Little Pigs*
Enzensberger, Hans Magnus, *The Number Devil*
Neuschwander, Cindy, *Sir Cumference and the Dragon of Pi*
Schmandt-Besserat, Denise, *The History of Counting*
Glass, Julie, *The Fly on the Ceiling: A Math Myth* (Grades 2–3)
Glass, Julie, *Counting Sheep* (preschool–Grade 1)
Turner, Priscilla, *Among the Odds and Evens*
Harshman, Marc, *Sólo Uno*
Juster and Feiffer, *The Phantom Tollbooth*
Abbott, Edwin, *Flatland*
Carroll, Lewis, *Alice in Wonderland*
Carroll, Lewis, *Through the Looking Glass*
Scieszka, Jon, Lane Smith, *Math Curse*

Family activities and resources for parents:
Barron, Marlene, *Ready, Set, Count*
Kenda, Margaret and Williams, Phyllis, *Math Wizardry for Kids*
Pappas, Theoni, *The Adventures of Penrose the Mathematical Cat*
Kenschaft, Patricia Clark, *Math Power*
Ma, Liping, *Knowing and Teaching Elementary Mathematics*

GAMES

Set: the family game of visual perception; 24 game; Mancala, chess, checkers;
Card games
Cluefinders Software

VIDEOS

Donald in Mathmagic Land
Schoolhouse Rock: Multiplication Rock

Figure 4.9.6. A list of resources for families.

Some Successful Tasks

1) "Area Race" (for kindergarten): The parents and children received 16 two-colored squares, red and green. They placed the squares on the table, forming a larger square, with half of the small squares showing red, and half showing green. The parent and the child each chose a color and alternated rolling a die. After each roll of the die, the player turned over that number of squares to show his/her color. The goal was to end up with just one color. Throughout the game, after each roll, each child had the task of filling out a table. The table asked for the number of red squares showing, the number of green squares showing, and the sum of the red squares and green squares. I was at the table a couple times when a kindergartener suddenly realized: "It's always 16!" The children also had the opportunity to rearrange the squares to form different shapes. But they found that in every case, the sum of red squares and green squares always resulted in 16.

2) "Roll Two Dice," adapted from *Math by All Means: Probability, Grades 3–4* (for second grade): Each student and parent rolled two different-colored pairs of dice and recorded their sum on a sheet under the appropriate number, from 2 through 12. For example, the roll of 2 and 5 would be recorded as 2 + 5 (the number on the red die was always recorded first). The child and the parent repeated this process until one of the columns got completely filled-that is, the number reached the finish line. Before the activity took place, the families had the opportunity to guess which number would "win." After reaching the end, all the children and parents compared their record sheets and had a discussion about probability.

3) "Tantalizing Tessellations" (for fourth grade): The future teachers showed the families some prints by M.C. Escher and talked about tessellations. Then the parents and children received different construction paper shapes (squares, triangles, pentagons, hexagons, circles, etc.). They made predictions about whether the shapes would tessellate the plane, and then experimented with them. The children made conjectures about types of shapes that would tessellate and those that would not, and then they made their own tessellating shapes "a la Escher."

4) "Estimatng and Calculating" (for several graders, each activity was at a different level, depending on the grade): The children and their parents took part in different activities estimating and calculating the number of items in different bags: beans, buttons, and the amount in different containers rice, water, the number of people in the room, the number of small and large feet needed to cover an outlined area of a parent on a piece of paper, the perimeter of the drawing, the number of tiles on the ceiling, etc. The children and parents also estimated how much water a sponge would hold, how far they could throw a paper plate, how many beans they could hold in their hand, etc. Another activity focused on estimating prices of different household items.

5) "How Much is a Dollar" (for the younger grades): The activity focused on all the different ways to make one dollar using coins. The children and their parents made a prediction about the number of ways to make a dollar. Then they set out to work, trying to come up with a solution.

6) "Platonic Solids" (for the older grades): This activity introduced the Platonic solids, and some terminology used to describe them: face, edge, vertex. The parents and children built their own regular polyhedra using toothpicks and small balls of clay. Then they recorded the number of faces, vertices and edges for each one on a worksheet. The families were then instructed to make observations about the recorded data, and to find patterns. Some parents and children "discovered" Euler's formula for these polyhedra and explored the concept of duality. The families also had the opportunity to create nets for the Platonic solids.

Figure 4.9.7. Some examples of successful tasks.

children select a book or software program from the display table. The evening activities conclude after the student leaders thank the attendees, give them a list of resources (see Figure 4.9.6) and ask them to fill out an evaluation form.

Some examples of successful tasks developed or adapted by the future teachers are shown in Figure 4.9.7. I consider an activity successful if it accomplishes the following:

1) It helps the families view mathematics as more than just a collection of arithmetic and geometric procedures and formulas to memorize—it gives them a glimpse of the type of work done by mathematicians. That is, the families engage in mathematical thinking, problem-solving, and dialogue.

2) It gives students the opportunity to experience an "Aha!" moment.

3) It entices the children and parents to continue exploring at home.

4) It gives the teachers some flexibility—they can adjust the activity by starting with more accessible problems, when necessary, as well as providing extensions for children who are ready.

The Aftermath

And finally, the Aftermath Phase is very important to document what went well, what lessons were learned, and how the next event can be improved. The following class meeting we spend some time reflecting on the events that transpired during Family Math Night. Approximately a week later, the students turn in reports including a description of the development process of their activity, their experience during the math family night, lessons learned, and their reflections.

A Look Back

As I now reflect on the several Family Math Nights that I have led my classes through, I can remember (and need to point out) that everything did not always go exactly as anticipated. I continue to receive feedback from the attendees, my students, and others, and I learn from it. For example, in the early events, we had too many families at one table and not enough at another. We have solved that problem by placing a representative from each grade at the entrance and having that person distribute the families among the groups.

Another area that was initially a challenge was finding the best way to set up the rooms to achieve efficient use of the space and easy traffic flow. We now work with kindergarten children and their families in classrooms. The children in the classrooms are not distracted by the noise and activity in the larger room, and the rest of the grades have more space available. A challenge we continue to have is encouraging enough of the families in the older grades to attend. At the last Family Math Night we had three activity booths set up for children in grades 6th through 8th (Longfellow goes up to 8th grade). Only two children in those grades showed up, while we had five full booths in kindergarten and first grade. We made an effort to "recruit" participants of all ages for those activities, and those particular future teachers had the added challenge of working with a large range of ages, from very small children to adults. They appreciated that opportunity, but we would like to come up with better ways to encourage the older children to attend.

Another challenge we have is that inevitably some of the future teachers are not able to attend the scheduled Family Math Night due to other commitments. We have tried a couple of solutions. One semester we held an additional Saturday Family Math Day at a different Elementary School. Another semester we held a mini-Math Family Session in an elementary school classroom during the school hours (without parents). The solution to these challenges will depend upon the facilities and equipment that are available. My point, however, is that I solicit feedback in the form of a "Satisfaction Survey" from all participants. I learn a lot from each of the responses and try to improve the event with each iteration. I also collect comments on these surveys and in the reflection assignments of the students. Here are some of these comments:

From the Future Teachers:

"I liked watching the children think. Sometimes a child would solve a problem faster than his parent. It is amazing to watch how some parents are involved—their children stay involved. Some parents would give up on the problem and their children did too. Attitude makes a big difference."

"It was amazing to see the concentration on the children's faces and their joy of doing math, for the pure fun of it. No reward other than accomplishment was necessary"

"It was neat when the parents, or even the students, would point out yet another way to solve one of the problems. This showed the students that there is not always "only one way" to get to an answer…This concept is important for children to see and learn, especially at a younger age. If they learn multiple ways to find an answer, then they have more in their mathematical repertoire to choose from."

"We would love to be a part of another math family night and even consider holding one at the schools where we will eventually teach."

"Both the students and the parents had very optimistic attitudes and were excited to be there. It was great to see so many families and children that were eager to work on math problems."

"Our experience at Math Family Night was not what we had expected but it was extremely worthwhile. As in life, nothing goes as planned and this was a wonderful way to force us to accommodate to our surroundings and to our students. As future teachers, we need to realize that our days are never going to be as planned and we need to be able to be flexible in changing the little details of our planned activity. "

"The Math Family Night was a good experience for us, as we were basically taught by the children." For instance, they allowed us to experience teaching one concept in many different ways. We also found out that there is a variety of levels in a classroom, as we were faced with children who needed to be guided, as well as children with advanced math knowledge."

From the Parents and Elementary Students:

"My first-grader and I attended Math Family Night and had a wonderful time. It was well worth the effort to get there on time. The USD students did a great job designing their stations, and my daughter thoroughly enjoyed herself."

"At math family night I realized that math can be fun!"

"Thanks to you and all the USD students for putting on a great event. It was an enjoyable evening for my two daughters and me. It has reminded me to spend some time with my oldest daughter, doing math together at home. The event has made her more receptive in the past couple of days to doing math with me at home. An event like Math Family Night demonstrates to her that math is highly regarded enough to create a special event for it. Thanks again."

"I learned humility at the Math family night. The activities challenged me to think a different way and they allowed my children and me to work together. We were entertained and intellectually engaged at the same time. I'm looking forward to the next one!"

The Road Ahead

The Mathematics Department and the School of Education have been actively pursuing opportunities for collaboration in order to enhance the educational opportunities for our future teachers. Next year we look forward to cooperating with professors in the School of Education to make Math Family Night an even better experience for our students. The students in the mathematics concepts course will collaborate with students in the mathematics methods course to create and implement their activities.

I look forward to holding the next Family Math Nights. I find them very rewarding and worthwhile. I hope this explanation has inspired you to try your own Family Math Night. It is very important for us to increase the opportunities around the country for positive math experiences. We owe it to our future teachers, who will in turn pass it on to our children. Please feel free to contact me with questions and good ideas for more successful Family Math Nights.

About the Author

Perla Lahana Myers is an assistant professor of mathematics at University of San Diego. She is passionate about teaching mathematics and especially enjoys working with future elementary school teachers.

4.10

The Community Math Teaching Project

Jennifer Morse and Joshua Sabloff

University of Miami and Haverford College

Abstract. The Community Math Teaching Project is a service-learning course in which undergraduates learn principles of pedagogy and gain a new understanding of geometry through teaching weekly labs to small groups of urban high school students. In this section, we discuss our objectives for the undergraduates and high school students; the structure of our class, including the mathematical and pedagogical content as well as a day-to-day schedule; and the extent to which our objectives were met.

Introduction

The Community Math Teaching Project is a service-learning course that stemmed from an initiative whose broad goal was to increase participation and achievement in science within the local school district. We began developing our course in the fall of 2001, structuring it around the idea of involving undergraduate students with teaching interactive geometry projects at a local high school.[1] We taught pedagogy and geometry to the undergraduates, with weekly sessions allowing them to implement what they learned with small groups of high school students. The hope was that the undergraduates would tailor hands-on activities to meet the high school students' needs, which ranged from reviewing basic skills to motivating geometric concepts through their applications; this would promote both the undergraduates' and the high school students' learning. We offered our class through the math department and the Center for Community Partnerships at the University of Pennsylvania in the spring semesters of 2002 and 2003. The course attracted students of all levels and in majors ranging from history to math; some wanted the opportunity to try out teaching as a possible career path, and others were drawn in by a spirit of volunteerism. Many of the students were in non-technical majors but were excited by mathematics. This

---

[1] Our course was loosely modeled after a project implemented in the biology department at the University of Pennsylvania.

Table 4.10.1. Roles and responsibilities of those involved in the class.

Professor (and TA)	Coordinate sessions among undergraduates and high schoolers. Teach undergraduates geometry and pedagogy. Design labs for use at the beginning of the semester.
High School Teacher	Help organize student groups. Encourage students to participate. Incorporate labs into the curriculum.
Undergraduate Students	Learn and implement pedagogical ideas. Learn and teach geometry. Create a lab for use at the end of the semester.
High School Students	Participate in weekly interactive geometry projects. Learn to work with classmates.

class afforded them the opportunity to learn mathematics in a context different from the standard introductory calculus sequence.

As a result of our positive experience—and despite conditions that varied from one year to the next such as class size, professor, high school curriculum, etc.—we are hopeful that the idea can be modified or adopted at other institutions. In this section, we will provide a framework for anyone who may be interested in running a similar course. The roles and responsibilities of the people centrally involved in our project are summarized in Table 4.10.1.

Objectives and General Course Structure

We divided our objectives for the undergraduates into two components:

Learn Principles of Pedagogy. We planned to cover a variety of approaches to teaching and expected the undergraduates to test the effectiveness of different methods when working with the high school students. We placed particular importance on how to assess and adjust to a student's understanding, and emphasized motivating students by revealing applications for the material or connections to their usual lessons. We also set out to develop an awareness of the high school environment, since this would play a central role in the effectiveness of certain techniques.

(Re)learn geometry. We anticipated that there would be a significant time lapse since the undergraduates had last studied geometry, as well as a wide variation in their background. Thus, we hoped to reinforce elementary concepts, to extend the basics to modern "transformation" geometry, and to explore applications in everyday life.

We endeavored to place our undergraduates in a situation in which they could best apply what they learned about pedagogy and geometry and in which their efforts could have maximum impact. Based on research articles and discussions with high school teachers, we identified the principal obstacles to academic success that we felt could be addressed within the limitations of our project. Since class sizes were beyond maximum capacity and students were arbitrarily shuffled between classes, the pace of the course could not meet the students' needs: those who were misplaced in the class or joined midterm struggled to follow, while others found the speed excruciatingly slow. These issues

Table 4.10.2. Overall Structure of the Course.

	Monday	**Wednesday**	**Assignments**
Introductory Segment (2–3 weeks)	Principles of pedagogy Geometry review Context in which they will teach		Pedagogical readings Journals
Weekly Labs (10–12 weeks)	Discussion/"debriefing" Lab preparation Summary of lab	Do the lab!	Readings from geometry text Journals Create a lab

gave rise to low attendance and difficulties in getting students to engage in classroom activities. To address these issues, we hoped to:

- Establish an environment in which small groups of students would be encouraged to refine their math communication skills and work together to learn. The personal attention from the undergraduates would be aimed at meeting students' individual needs, while the undergraduates would also seek to foster lasting academic peer relations among the high school students themselves.

- Work closely with the teacher's course plan, emphasizing topics with which students usually struggle and highlighting real-life applications of these topics. We hoped that our approach would reinforce the course material while motivating students to attend and actively participate in class.

It was natural to split the course into two weekly sessions, one devoted to the undergraduates and the other in the high school itself. Each Monday, we met with the undergraduates at the university, discussing problems they faced while teaching, the upcoming lab, and their assignments. On Wednesdays, we met at the high school to teach the labs. Within this framework, we shall discuss here the three main aspects of our course: an introductory segment, the labs themselves, and the assignments. In addition, we will outline some of the logistical preparations necessary to run the course. Table 4.10.2 summarizes the overall course structure.

Introductory Segment

We dedicated the first weeks of the semester solely to the undergraduates, preparing them for anticipated mathematical, pedagogical, and social issues in the high school. Each of these aspects is discussed below.

Review of Geometry

A review of basic definitions and theorems of Euclidean geometry was needed since the undergraduates had not studied high school geometry in as much as seven years. One undergraduate expressed the shortcomings of his high school learning process:

> "I found myself being able to do most of the problems, but I did not fully understand the concepts behind them. This superficial understanding was good for homework and quizzes, but rather bad for exams, where I suffered."

We placed special emphasis on viewing geometry as the study of properties of shapes that are invariant under rigid or affine transformations of the plane, rather than as a study of

Euclid's axioms and their logical consequences. This allowed the undergraduates to recall old knowledge in a new context and to focus on a conceptual understanding rather than subscribing to a procedural method. Part of our approach included heightening awareness of standards of argument: what does it mean to prove something? Techniques used during the review included:

- Breaking undergraduates into small groups to solve problems of moderate difficulty, such as finding the center of a rotation that moved a given triangle onto another. This tactic enabled us to integrate the geometry review with an introduction to pedagogical ideas (see below).
- Class discussions about basic definitions and theorems. For example, we led a discussion analyzing the structure and content of the introductory chapter of the high school textbook.
- Solving suggested exercises from the high school textbook.

Principles of Pedagogy

Our introduction to the principles of pedagogy centered around three questions.

What does it mean to understand a piece of mathematics? Through readings and class discussion, we explored the difference between procedural and relational understanding, the importance of contextualized knowledge, and the role of the ability to transfer knowledge from one setting to another.[2]

How can a teacher gauge a student's understanding? The undergraduates had a perfect opportunity to probe their students' understanding and to adjust their teaching accordingly during their weekly sessions. To prepare them to take advantage of this set-up, we used "jigsaw" activities, during which the undergraduates were split into two sets of small groups. Each set worked on a different geometry problem. We then formed new groups with at least one member from each set and asked our students to explain their problem to their new partners, paying careful attention to when and how they "knew" their partner understood the problem.

How can knowledge of a student's understanding be used to improve instruction? Here, we discussed motivational methods, analyzed types of questions to ask, and encouraged flexibility. As the semester progressed, we used episodes from the high school classroom to bring up these and related issues during weekly "debriefing" sessions. For example, after a lab in which the high school students were confronted with the question, "What is the perimeter of a shape with eighteen sides, each of length 1/4?", the students engaged in a discussion about the varying nature of their students' problems, shared the methods they used to deal with these difficulties, and suggested alternative approaches to their peers.

Context for Teaching in an Urban High School

Our weekly sessions with high schoolers were held in a large public school in the Philadelphia school district, enrolling more than 2000 students. The district averages a

---

[2] A good place to start exploring these ideas is *How People Learn* [1].

47% high school graduation rate after 6 years and of these students, only one in five has enough credits to even consider college. Therefore, tactics such as grade threats may not be effective. On the other hand, making the material relevant and interesting was crucial.[3]

Since the vast majority of our undergraduates came from suburban or private schools, we invited a number of speakers to discuss the environment at the high school and to provide some general insight into teaching in urban schools. The speakers warned that the level of competency with the course material would not reflect the student potential. This led to an ongoing mantra in the course: always underestimate background knowledge; never underestimate ability.

We also invited the high school teachers to speak more specifically about the students in the geometry class. They discussed logistics such as the class enrollment (30–35 students) versus attendance on a given day (15–20). The teachers gave anecdotal stories to provide a feeling for the class dynamic, explaining that the undergraduates may be asked on dates or offered frank opinions about their bad breath and weight gain. They sternly warned that the high school students were very sensitive and would be personally offended if the undergraduates did not show up.

Weekly Labs

When the semester started at the high school, we settled our students into a rhythm of discussing and preparing the lab during Monday's class and going to the high school to teach on Wednesday.

Monday

We broke our Monday classes into three segments:

Discussion of the previous week's experience in the high school. We brought up or solicited examples of successful and unsuccessful techniques. As the undergraduates became more comfortable with each other, this turned into a teaching workshop. We injected tips for teaching into these discussions, such as how to connect topics from previous labs and how to ask questions that encourage active learning.

Lab preparation. We broke the undergraduates into small groups to prepare the lab while we circulated through the class to answer questions and provoke discussions about specific points in the lab. We pushed them to think about ways to adapt the activities to meet their students' learning needs: some students needed more background support while others appreciated a good challenge. Lab preparation was also a good time to ensure that the undergraduates had a thorough understanding of the underlying mathematics. Finally, this is when we handed out materials.

Summary discussion. At the end of class, we asked the undergraduates to share their predictions about where their students may run into difficulty. If time permitted, we gave an enrichment lesson. For example, after preparing for a lab that used a fractal construction to show that the perimeter of a shape can grow arbitrarily large even if area is held constant, we gave a quick introduction to the idea of fractal dimension.

---

[3] For more information, see [4] and [6].

Wednesday

The first lab was designed to break the ice between the undergraduates and high schoolers, intending to set a fun, interactive precedent for the rest of the semester. One year, we asked the undergraduates to search the web for photos relating to a list of interests supplied by the high school students—basketball heroes, actors, cars, etc. The lab then included an activity for the students to decorate three ring binders with the photos, following certain steps relating to congruence theorems. The students used the binders for the rest of the year to hold future labs.

Subsequent Wednesdays started with the task of getting everyone settled into their groups and ensuring that they had sufficient materials for the lab at hand. Given the high absentee rate, we often had to re-assign groups, although we tried to keep the groups as stable as possible. Once the students were involved in the activities, we took on a background role so as not to undermine the authority of our undergraduates. We provided teaching support when necessary, but primarily made observations to bring up during Monday discussions.

Further information on the nature of the labs is provided later in this essay.

Assignments and Reflection

The undergraduates were given three types of assignments to enable them to meet the objectives outlined above: weekly readings, weekly teaching journals, and a semester-long project that culminated in the creation of their own geometry labs for use with their students.

Readings

The undergraduates were given three types of weekly reading assignments. They were asked to read the week's lab and to generate ideas for testing and discussion during the lab preparation segment of the Monday classes. They were also required to read the relevant sections in the high school geometry textbook so that they would have a sense of what their students were learning in class and what material to reinforce. The last type of reading assignment came from articles and selections from books on teaching and learning mathematics. Some of the material that we found to be effective in raising questions and stimulating discussion included Skemp's article [5], which describes the difference between teaching for procedural understanding of mathematics and for a more conceptual and contextualized understanding; Ma's book [2], which raises the issue that both content and pedagogical knowledge are necessary for effective teaching; and the NCTM's *Principles and Standards for School Mathematics* [3], which gives the undergraduates an idea of what expectations can be set for their students.

Journals

The undergraduates maintained a weekly teaching journal based on their high school classroom experience, readings from homework, and in-class activities. We asked them to reflect on several questions including: what were their student's preconceptions about the material, what techniques did they try and were they successful, and how were ideas from class and the readings incorporated into their teaching? We also asked the undergraduates to discuss how their own understanding of geometry changed (if at all) during the semester, and to comment on the specific mathematics in the labs. They addressed all these questions, though they focused largely on their experiences teaching unless explicitly asked to do otherwise. The open-ended nature of the journal assignments led to many interesting responses. For example, one entry read:

> "I found myself doing a lot of work for the kids in the beginning and moving on without knowing if they had a strong comprehension of what was going on. I realized this mistake halfway through the class, and attempted to adjust my role."

Creation and Implementation of Labs

A large component of the course was to design a lab that the class could implement with the high school students. Each group of two to four undergraduates was asked to create a series of activities and worksheets that would last 45 minutes. We expected their labs to give an indication of how well they were able to assess and identify shortcomings in prior labs, to accommodate students with varied ability and knowledge, and to motivate the course material. As part of the final product, the group had to turn in a short explanation of the choices made in developing the lab, linked to class discussions, the readings, and experiences in the high school. We established the following guidelines for the lab:

Motivation. The undergraduates were challenged to investigate how their topic arises in art, nature, or pop culture, and then to incorporate their findings into their labs. For example, the undergraduates took photos of the students and then stretched them in various ways to demonstrate how changing the ratio of sides distorted their faces.

Multiple approaches to one topic. We emphasized that the labs should focus on one or two points and provide several activities pertaining to the same idea. For example, a lab on similarity and ratios consisted of three parts. First the students experimented with changing the lengths of sides of photos to observe how similarity is the only way to avoid distortion in enlargements. This was followed by an activity to form various similar polygons using rubber bands and geoboards. The lab ended by using proportionality to make trail mix for the group, given the recipe for ten people.

Use simple, spare instructions. We found that requiring the undergraduates to avoid paragraphs of text assured that they would give simple explanations for the mathematics and that their activities would be explained clearly and effectively. This was further necessitated by the fact that some of the high school students were constrained by their poor reading skills. The need to make the labs adaptable to different student levels was also met more easily with short, diverse activities. We encouraged the undergraduates to include preliminary sections devoted to teaching the students definitions by example. In a lab about Eulerian paths in graphs, for example, the undergraduates introduced the idea of a graph using a map with roads as edges and towns as vertices.

Ask for explanations. The high school students' math communication skills were enhanced by exercises that required them to write down their ideas and explanations.

- The actual creation of the lab was broken into several assignments:
- Outline the concepts to be presented and goals for the lab.
- Brainstorm about possible activities and applications, including a list of necessary resources.

- Write a rough draft of the lab, including a list of necessary manipulatives, paying close attention to focusing on main points and motivating them.
- Finalize the lab and test run it with the group, revising as necessary. This requirement helped to catch vague questions and unfocused activities.

Logistics

A significant amount of preparation is necessary to make this class run smoothly. In addition to the usual preliminaries (setting a syllabus, choosing readings, etc.), labs need to be created, schedules need to be coordinated, and the class needs to be advertised in order to enroll at least half as many undergraduates as high school students.

A good working relationship with the high school teacher contributes greatly to the success of the class. To be able to integrate the labs into the teacher's lessons, a good estimate about the speed of progression through the curriculum is necessary. Further, the teachers can point out areas where their students need extra support, thus enabling us to design the labs that would place the undergraduates in a situation where they can address the high school students' individual needs. Using this information, we created the labs based on the criteria discussed in the previous section. Note that this was a time-consuming process and merited a course reduction at the start of the project. So long as the high school curriculum remains consistent, the labs can be reused from year to year.

Certain technicalities must also be attended to: for example, with the teacher's help, we set a schedule with the fewest number of in-service days and holidays to minimize interruption of the project. In addition, a classroom in the high school that accommodates a doubled class size (or two smaller rooms) is also needed.

Impacts and Challenges

Naturally, it is difficult to measure precisely the impact of such a project; in fact, we admittedly have no statistical information about the long-term effects of this project on either the high school or undergraduate students. Nevertheless, we shall try to pinpoint areas of perceived success and discuss aspects of the project that posed difficulties.

One undergraduate summarized the experience as follows: "I learned so much about *learning* in this class." Comments such as this indicate that the class provided an opportunity for the undergraduates to become self-aware learners. We gave the students multiple opportunities to reflect upon their experiences: individually with weekly journals, in small groups during the process of their lab construction, and as a class during weekly debriefings. The high school spring break fell toward the end of the university semester, providing the undergraduates with a chance to digest their experience together in a wide-ranging discussion. We touched on what they learned about geometry and teaching, emphasizing their reaction to unfamiliar sets of goals and expectations, poor mathematical backgrounds, and low frustration thresholds. The undergraduates also grappled with topics such as the university's responsibility to the community, the efficacy and consequences of tracking, and the role of standardized testing.

With respect to our mathematical goals, many of the undergraduates commented that they were motivated to gain a deep understanding of the material in order to be effective teachers. This is especially gratifying after many of them had used their first journal entry to bemoan the inadequacy of their own high school geometry experiences. Their students kept them on their toes by asking questions that they did not anticipate: the undergraduates were asked intuitive questions when they expected a deductive line of reasoning to suffice. For example, one undergraduate wrote,

"Even when I provided them with examples that I thought were confusing, or had confused me on Monday during our practice 'run,' they had the right answers. I think that their visual orientation and amazing artistic abilities helped them tremendously in their conceptualization of the shapes and their characteristics."

Our classroom observations revealed that the undergraduates' understanding of what constitutes a valid argument could be improved. While there are situations when mathematical standards of validity can be relaxed, we wanted to be sure that the undergraduates were aware of what was being swept under the rug. Furthermore, by replacing a few of the debriefing sessions, there is room in this course to explore geometry more extensively.

Overall, the students found teaching the geometry labs to be more challenging than they expected at the start of the course. Their own experiences in exceptional schools with high importance placed on education are likely to have contributed to this. The most striking difference and the one posing the most difficulty was an insufficient background in the high school students' knowledge. One undergraduate noted:

"I realize how integral a part of math geometry really is.... It depends on lessons learned in previous courses and is virtually impossible when the student is lacking other basic math skills."

The undergraduates developed several strategies for dealing with these gaps, from specifically addressing problem areas at the beginning of the sessions to changing the emphasis of a given activity. Such adjustments made by the undergraduates also contributed to the success of meeting the high school students' educational needs. For example, one undergraduate found that his students loved "bonus" problems, and each week he prepared an extra exercise for them. As they gained confidence and patience over the course of the semester, many of the undergraduates' questions began to *lead* rather than ask for simple information or implicitly *tell*.

Our undergraduates quickly recognized the importance of finding a hook to grab their students' attention. Their methods varied from convincing their students that the mathematics itself was interesting to connecting the labs to pop culture or everyday experience. For example, one group of undergraduates wrote about the lab they designed, "We liked that the concepts could be rooted in real-life situations ... which would provide motivation for students and answer the proverbial question, 'when will we ever use this in life?'"

This would not have been a worthwhile service-learning class if the high school students had not also benefited from the efforts of the undergraduates. Certain accomplishments in the high school classroom were easy to gauge. Attendance at our Wednesday labs averaged 25 students in comparison to 15–20 on other days, though this can also be attributed to the fact that one year the teacher made a special effort to beg and threaten her students to come on Wednesdays. The immediate task of engaging students in the math labs was another obvious success. The classroom buzzed with students absorbed in the labs or discussing (and sometimes arguing about) math. Often students skipping other classes wandered by the classroom and after peering in, begged to join us. Teachers stopped by to marvel at the attentiveness of the high schoolers during these sessions.

Given students of such disparate mathematical backgrounds, many of our goals were attained as a direct consequence of the small group setting. The students themselves were acutely aware of the overcrowding problem that faces their school, commenting that:

> "They helped me learn more because you had just back to back help instead of just one teacher for ... 33 students."

> "They give us time to learn. They don't move on until we understand the problem."[4]

Our lessons motivated the subject by revealing math in the students' environment. For example, the students practiced measurement skills and analyzed scale factors by drawing maps of their classroom. After arranging an outdoor scavenger hunt during which the students were asked to identify certain isometries such as those of car emblems, our efforts were rewarded with comments such as, "Whoa, tessellations are everywhere!" A short illustrative excerpt from a basketball lab is included in Figure 4.10.1.

In light of the small student-teacher ratio, one of our primary objectives was to foster an academic relationship among the high school students that would extend beyond our sessions. We made every effort to encourage the students to go to each other with questions instead of waiting for the overextended teacher. We arranged the desks so that high schoolers sat side-by-side, facing the undergraduates, with the idea that it would more natural to turn to each other for help and to work together. Labs often involved interdependent activities designed for more than one student, such as playing "go fish," with matching cards defined by similar triangles. When appropriate, the students would take turns reading questions aloud and were often asked to explain concepts to one another in their own words. This tactic was not only an effective reminder that the students could help one another ("Some things the teachers couldn't explain, they could") but it also refined the students' mathematical communication skills. Despite difficulties with finding compatible groups and fluctuating attendance, we were extremely pleased when we visited the high school class after Penn's semester ended and found the students working on a worksheet in many of the same groups that were formed during our project.

Though, as has been mentioned, we do not have any hard data about the long-term effects of this course on the high school students' learning, we believe that the positive anecdotal evidence outlined above is sufficient to justify a continuation of the project, especially when this evidence is combined with the strides that the undergraduates made as both teachers and learners.

References

1. J.D. Bransford et al., eds., *How People Learn,* National Academy Press, Washington, DC, 2000.
2. L. Ma, *Knowing and Teaching Elementary Mathematics*, Lawrence Erlbaum Associates, Mahwah, NJ, 1999.
3. National Council of Teachers of Mathematics, *Principles and Standards for School Mathematics,* NCTM, Reston, VA, 2000.
4. G. Seiler, K. Tobin, J. Sokolio, "Design, Technology, and Science: Sites for Learning, Resistance, and Social Reproduction in Urban Schools," *Journal of Research in Science Teaching*, 38 (2001), 1–22.

---

[4] These, and other quotes in this section, were taken from anonymous written feedback from the high school students.

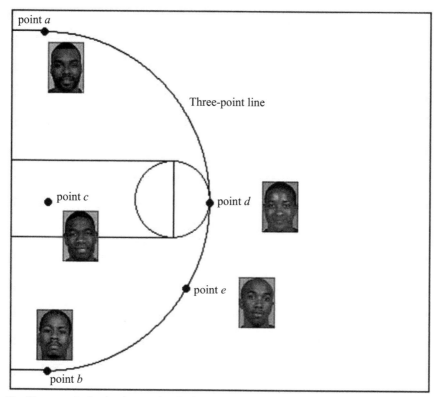

The Sixers are losing by three points in the final minute and desperately need a basket. Allen Iverson has the ball at point b, but is being double-teamed and can't shoot. Aaron McKie is at one point wide open to take the shot, but Iverson can't pass to McKie directly.

Iverson has two alternatives:

1. He can pass to Eric Snow at point d, who then passes to McKie at point a. Draw a line connecting b to d and d to a (symbolizing this pass)
2. He can pass to Speedy Claxton at point e, who then passes to McKie. Draw a line connecting b to e and e to a (symbolizing this pass)

For Snow or Claxton to pass as fast as they can, they want to pivot as little as possible before passing the ball. Assume the amount they turn is equal to the angle measure at the point where they are standing. Does it matter who Iverson passes to? Hint: consider the theorems used in your last activity.

Figure 4.10.1. Problem from the angle measure and basketball lab.

5. R. Skemp, "Understanding Relational Instrumental Theory," *Mathematics Teaching*, 77 (1976), 20–26.
6. K. Tobin, G. Seiler, and E. Walls, "Reproduction of Social Class in the Teaching and Learning of Science in Urban High Schools," *Research in Science Education*, 29 (1999), 171–187.

About the Authors

Jennifer Morse is an assistant professor in the Department of Mathematics at the University of Miami. **Joshua Sabloff** is an assistant professor in the Department of Mathematics at Haverford College. Both authors became interested in service learning while teaching at the University of Pennsylvania.

Chapter 5

Untapped Possibilities?

Charles R. Hadlock
Bentley College

Introduction

As was discussed in Chapter 1, I began this project by conducting two national surveys of service-learning projects in the mathematical sciences, one targeted at mathematics faculty members and the other at service-learning professionals. At that time I had no idea how I was going to organize the results, but it soon became clear that they fell relatively neatly into the subject areas represented by the three previous chapters of this book, namely, modeling, statistics, and education. While this seemed quite convenient from the standpoint of book organization, it was also somewhat disappointing. Were there not other areas of the mathematical sciences in which service-learning might play a valuable educational role? In fact, I had already thought of some other areas in which I had expected to encounter projects.

Therefore, I decided to review the entire scope of undergraduate mathematics courses with a view to identifying additional service-learning possibilities. This process began with Table 5.1, which shows the typical range of undergraduate mathematical sciences courses. Few schools are likely to offer this entire range, but that is not important. The fact is that most schools will offer courses at various levels, ranging from remedial or beginning college level to much more advanced; these courses will usually represent a blend of both pure and applied mathematics, and in addition there may be courses specifically targeted to future elementary and secondary school teachers. While the mathematical reader is no doubt already familiar with the content of Table 5.1, it may be useful to keep this table in mind as I try to suggest how further service-learning opportunities may be lurking throughout the curriculum. Perhaps my own suggestions in combination with this table may stimulate further ideas on the part of the reader.

In my surveys, relatively few of the courses shown in Table 5.1 surfaced as examples of courses with integrated service-learning projects. The reader may be inclined to dismiss relatively quickly some of the advanced courses in pure mathematics as candidates for service-learning, but in a moment I will ask that this be given a second look. However, most striking to me in reading the survey results was the absence of projects related to the time value of money (for example, net present value), which is a fundamental topic in business calculus courses and courses in the mathematics of finance, and which is also likely to arise in college algebra, precalculus mathematics, and probably elsewhere. Therefore, in comparing the survey results with the course list in Figure 5.1, I began to formulate a four-part thesis as the basis for this chapter, namely:

1. There is potential for many service-learning projects in the areas of financial mathematics and organizational management.

Table 5.1. Typical range of undergraduate mathematical sciences courses (classifications may vary somewhat with institutional context.

Classification	Typical course title
Elementary	remedial mathematics college algebra finite mathematics mathematics for the liberal arts precalculus trigonometry calculus (many varieties) mathematics of finance basic statistics
Intermediate/Advanced	linear algebra abstract algebra multivariable calculus differential equations real analysis complex analysis linear programming numerical analysis optimization methods operations research discrete mathematics advanced mathematical finance topics in actuarial science number theory probability mathematical statistics topology geometry (advanced) dynamical systems mathematical logic applied mathematics and modeling
Education	mathematical content for teachers methods and pedagogy

2. There are service-learning opportunities in advanced mathematics courses, including ones in pure mathematics.
3. There are a number of specialized opportunities involving the application of geometry and trigonometry that might be valuable and motivating to students studying these subjects.
4. There is considerable opportunity to increase the use of service-learning projects in courses in mathematical modeling and applied mathematics.

In the remainder of this short chapter I shall address each of these components in turn, attempting to illustrate it with hypothetical examples that I hope might suggest useful directions to the reader.

Financial Mathematics and Organizational Management

There is a relatively wide variety of introductory level college mathematics courses. Two fairly universal learning objectives for such courses are an incremental increase in the students' ability to handle quantitative methods and the development of a richer appreciation for the applicability of mathematics in modern society. Thus it is not surprising that issues associated with the time value of money are a topic represented in many of these introductory courses. It is of course an important topic from a real-world perspective because of the universality of bank interest, credit cards, mortgages, car loans, and other financial instruments with which the students probably already have some acquaintance. It is also of interest from a mathematical standpoint because it can be used to introduce sequences, limiting processes, and the exponential function. If the students are able to internalize this important concept, then it is likely to serve them well in understanding many practical financial issues that they are sure to encounter.

Table 5.2 presents some hypothetical examples of service-learning projects in the general areas of financial mathematics or organizational management. The first two project examples involve the time value of money. On the one hand, while corporate property managers are quite sophisticated in using this concept to manage decisions about capital investments, local government and community organizations are much more likely to overlook this kind of analysis in favor of short-term questions such as "Is the political climate right to get this purchase approved in this year's budget?" One wonders what the financial cost of this approach might be. On the other hand, at the level of the individual rather than the organization, a great deal of unnecessary debt is probably accrued because of an inadequate understanding of compound interest and time value of money issues. Thus it seems that this area is one where increased student practice would be valuable for the students and for those receiving the assistance. Typical courses into which such a service-learning project might be integrated could include college algebra, precalculus, business calculus, finite mathematics, and an elementary course in the mathematics of finance.

The third example suggested in Table 5.2, developing spreadsheets, may at first glance seem too simple, but I have seen its value in my own classes. In particular, the care and attention that students must give to algebraic expressions (especially relatively complex ones) when programming them into a spreadsheet formula is useful training, in my opinion, in developing habits of precision and accuracy. Many local agencies suspect that spreadsheet methods could help them accomplish some of their tasks more effectively, but they may not

Table 5.2. Potential service-learning opportunities in financial and organizational topics.

Representative project	Course/topic
Analyze capital expenditure plans for a not-for-profit organization; for example, optimal vehicle replacement schedule, buy vs. lease decisions.	College algebra, finite mathematics, calculus, basic mathematical finance. Time value of money, present value.
Support consumer advisory service by helping consumers understand the true cost of credit purchases and leases, perhaps through programs for special audiences (for example, welfare-to-work) or as part of an adult education program.	College algebra, finite mathematics, calculus, basic mathematical finance. Time value of money, present value.
Develop spreadsheet programs as needed to facilitate management within not-for-profit organizations.	College algebra, mathematics of finance. Careful development of algebraic expressions, order of operations, possibly the solution of simultaneous linear equations or a nonlinear equation in a single variable.
Recommend a voting method for a community organization, taking into account the organization's principles and values.	Finite mathematics/mathematics for the liberal arts. Alternative voting schemes, the measurement of power.
Carry out a critical path analysis for a complex project about to be undertaken by a local organization.	Finite mathematics. Critical path analysis, scheduling issues.

have the time or expertise on their staff to set up the spreadsheets with the underlying formulas. (They usually do have the expertise to use the spreadsheets once they have been constructed.) Such formulas always involve algebraic expressions; in addition, they sometimes involve the solution to algebraic equations. It can be an important experience for a student of college algebra or a related subject to see how such algebraic expressions are used to facilitate business management within a community agency and how the solution of simultaneous linear equations or a single nonlinear equation may also arise in practice.

Another group of elementary courses are those in the area of finite mathematics. This title is intended to contrast them with calculus-type courses that involve limiting processes. Finite mathematics courses generally cover a variety of topics chosen from, among others: probability, matrices, linear programming, voting systems, critical path analysis, game theory, and logic. The last two examples in Table 5.2 are based on two of these topics. With respect to the first, the topic of voting systems can actually lead quite far into the question of how different systems represent different divisions of power within an organization. It is easy to imagine how an organization might well be interested in choosing a system that might best align with its underlying philosophy. A well designed service-learning project might be the key to making this possible.

Of comparable value might be assistance from finite mathematics students to help an organization plan a complex project and estimate the criticality of various tasks in keeping the project on schedule. Depending on the nature of the project, it might be amenable to standard critical path analysis, or it might evolve into some related form of flowcharting or event tree construction. In any case, there is considerable value in exposing students to a real-world case where graphical and analytical techniques are useful in understanding the interdependence among components of a complex undertaking.

Advanced Courses Other Than in Modeling and Statistics

Shifting the focus now to a more advanced application of service-learning, I note from my survey results and from the examples provided in Chapter 4 that essentially all the education related applications of service-learning seem to be targeted at remedial situations, support in normal K–12 math courses, and perhaps some enrichment work at the elementary and middle school levels. What about the undiscovered or under-encouraged high ability high school student whose school district does not have sufficient resources to nurture his or her talent? I envision a situation where such students can meet and interact with talented university students studying advanced mathematics. For example, I remember an almost burning interest I had as a high school sophomore, after studying geometry, to understand how people could say that it was impossible to trisect an angle with compass and straightedge. But there were no teachers available with the background to guide me through this issue. At the same time, it might have been a marvelous experience for a student of field theory at a local university to reinterpret his or her pathway through that subject so that I could understand the trisection issue directly.

There is no shortage of fascinating topics in advanced mathematics that might open the door to an entirely new and interesting world for bright high school mathematics students. These topics cover the full range from number theory to topology to algebra to geometry and so on. Modern applications to cryptography, film special effects, computer games, and computer technology are equally captivating. How can we dangle this world in front of students who may not even know their own talents and who would not otherwise have the opportunity to encounter it? Here is where I think we have an untapped possibility for service-learning that can bear great benefit for all parties involved. Several such potential project approaches are suggested in Table 5.3, and there are of course many variations.

Are enterprises such as this really also of value to the advanced mathematics students in the university? I believe that the answer is definitely affirmative. Mastery of the theorem and proof style of many advanced courses does not necessarily provide the student with the opportunity to play in a more leisurely fashion with the concepts being discussed or to rework them from a different point of view. But for a student in a Galois theory class to try to explain to a high school sophomore why it is impossible to trisect an angle is likely to require a great deal of rethinking and reworking, in the end leading to a much richer understanding of Galois theory itself. After I went off to college I remember submitting problems to a local mathematics league that had sprung up in my hometown, and the benefits to me of this activity were certainly considerable and quite different from the experience of participating in regular mathematics courses. It gave me a different role and a more empowered attitude in my encounter with the subject, and in this way it complemented my course experiences. It also led to some interesting collaborations with student

Table 5.3. Potential service-learning opportunities in miscellaneous advanced courses, including ones in pure mathematics.

Representative project	Topics addressed
Run a speakers bureau for top students to speak to high school mathematics classes or mathematics clubs on catchy topics associated with the advanced fields being studied.	Subject matter of courses, reinterpreted at a level appropriate to audience.
Support local or regional mathematics competitions by suggesting problems, assisting grading, and in general organization.	Subject matter of courses, reinterpreted at a level appropriate to audience.
For schools that do not offer advanced placement calculus but where some students are interested in learning this material on their own or under the guidance of a teacher, provide support activities such as problem sessions, practice exams, and tutorials.	Review of basic calculus with the benefit of more mathematical maturity.
Organize at the university a math fair or math symposium targeted to gifted secondary students from multiple schools.	Topics driven by participants; challenge to university students is to understand the diverse subjects treated.

peers where we were not being driven at all by pressure from professors or course assignments.

Geometry and Trigonometry

Let me next mention a number of somewhat specialized service-learning projects related to the subjects of geometry and trigonometry, and perhaps therefore also relevant to students studying analytic geometry in precalculus courses. These are shown in Table 5.4, and they are largely self-explanatory. It is certainly well recognized that there are numerous applications of these subjects in the engineering fields, but it might be harder to imagine how students can gain practical experience and reinforcement of these concepts while providing a valuable service to the community. That's the point of the examples provided in the table, all of which are based on personal experiences of my own. Once again we see in the fourth example, calculating areas of irregular plots of land, the familiar theme of using the process of programming a spreadsheet to provide practice in the use of mathematical equations. Not only does this vehicle force students to be precise and to track down their own errors, as was mentioned before, but in the end it gives them a satisfying feeling of accomplishment for having used their mathematical formulas to produce a "product." I believe that modest successes such as this tend in the long run to build confidence and to improve attitudes towards mathematics.

Table 5.4. Potential service-learning opportunities in precalculus and trigonometry.

Representative project	Topics addressed
Collect GPS data in the field along a trail or other curvilinear feature, and use the data, perhaps via a simple spreadsheet program, to analyze steepness along the pathway (which introduces the issue of averaging distances); carry out in support of a hiking organization responsible for the maintenance of trails or a conservation organization concerned about soil erosion along skidder trails used in logging.	Slope, angle of elevation, average slope over various distances, graphical representation of results of analysis.
Participate in a volunteer project to renovate an older building or apartment and write up the various geometric and trigonometric issues encountered in planning cuts of wood or wallboard. (Older buildings invariably have many joints that are out of square.)	Approximating the tangent of an angle, for small angles, by the angle in radian measure; intersection of planes in the three-dimensional space (compound miter cuts); geometric shapes.
Research old deeds in support of a local historical society; produce plot plans.	Angle and distance representations, resolving inconsistencies.
Develop a spreadsheet program to calculate the area of an irregular parcel of land, based on a surveyor's description of the boundary lines; apply in support of a local historical society documenting old properties.	Triangulation; law of cosines; area of triangles.
Assist in surveying and/or map generation for af conservation organization.	Angles, distances, sources of error.

Additional Opportunities in Modeling

In Chapter 2, we saw several examples of successful modeling-oriented service-learning projects, and one can certainly imagine many additional variations. I have included in Table 5.5 a few more examples of modeling projects that might lead the reader to think in additional directions. As mathematicians, we certainly recognize that mathematical modeling of this type is used by many different kinds of organizations within our society, but it can still be challenging to find local organizations or government agencies who are willing to invest their time and resources so as to interact with our students on such projects. That's why three of the five examples in Table 5.5 are in the environmental field, which is full of issues amenable to modeling at a level accessible to average undergraduates and where there are almost always a number of involved parties, some of which usually lack the kinds of resources necessary to hire professional consultants.

Table 5.5. Potential service-learning opportunities in other applied areas

Representative project	Topics addressed
Review the utilization of traffic enforcement resources by a local police department and analyze the potential for more efficient deployment.	Model development, optimization techniques.
Review the environmental impact study of alternative plans for a local highway project and test various assumptions leading to the selection of a preferred alternative; client organization might be a local citizens group or town government organization	Quantitative figures of merit, data quality issues, sensitivity analysis (to assumptions).
Review the environmental impact study of alternative locations for a relatively undesirable fixed facility (such as a hazardous waste processing plant) and test various assumptions leading to the selection of the preferred alternative; investigate optimality methods for improving site selection; client organization might be a local citizens group or town government organization	Quantitative figures of merit, how to balance quantitative and qualitative constraints, sensitivity analysis, geographic information systems (GIS), utility theory and optimization.
Analyze almost any process within a university or other organizational environment with a view towards measuring its efficiency or optimality and/or proposing improvements. Typical examples might be course scheduling, registration, and local transportation.	Model development, quantitative figures of merit, sensitivity analysis, optimization techniques.
In conjunction with an environmental group, develop contour lines to represent an environmental variable of interest (for example, groundwater contamination levels, air pollution levels), based either on interpolation from data points or on finding the level curves from an analytic model. Use concepts from multivariable calculus to make deductions about the rate and direction of movement of contaminants.	Relation among directions of level curves, gradient vectors, and transport.

Conclusions

In closing this chapter, I should recognize that many of the service-learning possibilities I have characterized as "untapped" may actually have been pursued in some form in some institution. However, I certainly was struck by the fact that very few projects along the lines discussed in this chapter surfaced during the course of my surveys, and I firmly believe that there is substantial room for further utilization of service-learning, both to enhance the learning process itself and to render valuable community service in some of these areas.

About the Author

See Chapter 1 for information about the author, who is the Editor of this volume.

Chapter 6

Getting Down to Work—A "How-To" Guide for Designing and Teaching a Service-Learning Course

Jennifer Webster and Charles Vinsonhaler
Harvard University and University of Connecticut

Introduction

In this chapter we discuss resources for developing and supporting service-learning projects, advice for finding a community partner and developing a successful partnership, and guidelines for preparing students to work in the community. We also include a timeline for running a service-learning course, sample reflection questions, and a bibliography of resources on service-learning pedagogy, service-learning course design, and project assessment.

Defining the Concept of a Service-Learning Project

If you are intrigued by the concept of service-learning as outlined in Chapter 1, and inspired by the actual service-learning projects described by the authors of Chapters 2–5, then this chapter can help you get started in designing and teaching a course with a service-learning component. We primarily address the incorporation of service-learning into an existing course, but many of the concepts here are equally applicable to some of the other creative frameworks discussed in earlier sections, such as project centered courses and out-of-term experiences.

The first task is to determine the mathematical learning objectives for the course that you plan on teaching. Once you have determined these, you should begin to think of real-world situations where students can apply the mathematical concepts that they study in your course. Good service-learning projects for quantitative courses allow students to explore how mathematical models and ideas can be applied to solve real-life problems.

For example, faculty teaching mathematics courses for educators will look for community partner organizations where their students can be a part of teaching and tutoring programs that will help prepare them for their future teaching careers. Others may have a class that can use operations research concepts in order to improve the efficiency or effectiveness of some kind of process or operation. Instructors (or various community liaison persons within the college or university) can ask the town government if they are interested in assistance with tasks such as efficient driving routes for school transportation, snow plowing, street cleaning, and/or police cruisers. Keep an open mind as you talk to the agencies in the community, because a "ready-packaged" idea may not meet community needs. Also keep in mind that the project need not address all your learning objectives. Indeed, some of the authors of previous sections have used the service-learning project primarily to enhance the students' understanding of an applied field and, as a result, their commitment to learning the related mathematics presented in the classroom.

Resources Inside and Outside the University

If you are planning on integrating service-learning into a mathematical sciences course, there are almost surely internal resources at your university or college that can help you to design your course and to identify and contact an appropriate community partner. Many of the authors in this book comment that their project was particularly successful because they were able to work with a project coordinator provided by the service-learning office or financed by grant funding procured for service-learning programming. You should be aware of the fact that a nontrivial portion of Federal work study funds are required to be used to support community service, and thus there is likely to be much more technical or logistical support available to you than you might realize.

The first place to look for assistance is the service-learning office, if your campus has one. The service-learning staff can assist faculty with finding a community partner and can help with course design, reflection activities, and assessment. They are accustomed to working with faculty in many disciplines, and for an often underrepresented field like mathematics, they are likely to exert even extra effort to help it work. The service-learning staff can also help with the actual logistics of setting up and running a service project, and in some cases they can assign a work-study student to serve as a "course-assistant." If your campus does not have a service-learning office or designated service-learning coordinator, an on-campus sStudent community service center or student volunteer center can be a source of contacts with community partners. A volunteer or community service center may be more involved in direct-service programs for children, the homeless, or the elderly; however, the community organizations that provide these services may welcome more technical assistance to solve a financial or logistical problem. For example, the community service-center may have an established relationship with the local Boys and Girls Club, providing mentors or tutors for the children. By talking to the director of such an organization, you may be able to find a suitable mathematics-based project for your students, such as providing a statistical analysis for an annual report, a long range strategic plan, or a funding proposal.

Although the community service center may be a source of community partner contacts, it is important to keep in mind that community service is not the same thing as service-learning. Chapter 1 of this volume provides a discussion of the distinction between service-learning, community service, and other community-based learning experiences. On some campuses, the service-learning center and the community service center are merged, on some campuses they are separate, and some campuses have only a community service center. If you choose to seek help from the campus community service center, keep in mind that the staff may not have expertise in important areas of service-learning, such as defining the academic learning outcomes of the service-activity and incorporating reflection into service-learning-based coursework. In this case, you will have to rely on the expertise of colleagues working in service-learning, and on print and on-line resources such as those cited in the annotated bibliography at the end of this chapter.

If there is no service-learning office on your campus and the community service center cannot connect you with the type of community partner you are seeking, there are likely to be other offices on campus that can help. The Community Relations office can often provide contacts within appropriate local government and non-profit agencies. Offices charged with Experiential Learning or Excellence in Teaching may also be a good source of community partner contacts. Internship Coordinators, Cooperative Learning Programs, or the Office of Career Services may regularly place students at government agencies or non-profit agencies, or receive requests from such organizations. The Office of Sponsored Programs (grants office) may be able to help faculty identify and apply for grants to support a service-learning endeavor. This latter office is also often charged by the administration with finding novel initiatives to work into funding proposals, and they might particularly welcome some involvement in your planning efforts.

Academic institutions have used service-learning grant funds to hire course assistants, to cover incidental expenses such as transportation or tutoring materials, and to compensate community partners and faculty for time they spend on campus/community collaborations. Other possible sources of support are offices or committees charged with new initiatives on campus. For example, if your institution places an emphasis on utilizing course concepts outside of the classroom, community-based projects are a perfect match. If your institution places an emphasis on writing across the disciplines, the writing that service-learning students do as a part of their critical thinking about their service project fits in well with this institutional academic goal. If your institution emphasizes student civic engagement, service-learning courses are a natural fit since service-learning courses should be designed to engage students in public problem solving.

Other departments may already be partnering with local agencies or organizations on service-learning projects. An agency that already works with someone in your university may welcome quantitative assistance from mathematical sciences students and faculty. For example: students from liberal arts courses at Bentley College have completed service-learning projects with the children in the public schools in Waltham, MA for many years. However, in the spring of 2003, the City of Waltham and the Waltham Public Schools received more technical assistance from mathematics students in Charles Hadlock's interdisciplinary seminar. Waltham is in the process of tearing down old schools and building new ones. The mathematics students developed a model to analyze the costs and possible benefits of recycling building materials during this process. The partnership between the college and the town was already established, and the partnership was strengthened and given more breadth when town employees collaborated with mathematics students and faculty.

Table 6.1. Possible sources of institutional support for service-learning activities.

- Service-Learning Office
- Student community service or volunteer center
- Community relations office
- Offices charged with experiential learning or excellence in teaching
- Internship coordinators, cooperative learning program, or office of career services
- Office of sponsored programs (grant funding)
- University's current initiatives and strategic plans
- Other academic departments engaged in service-learning
- National Campus Compact office
- State Campus Compact offices

Table 6.1 provides a summary of suggested offices or departments that can be of assistance to faculty and students engaged in Service-Learning.

In addition to assistance in finding community partners, faculty engaged in service-learning often find that they need additional resources in order to efficiently set up their community-based projects and handle much of the logistics. Table 6.2 provides a list of possible sources of people-power on your campus.

If you have a service-learning office at your college or university, you can investigate what type of assistance the center offers to faculty who teach service-learning courses. Some colleges offer stipends, awards, or other forms of recognition to faculty engaged in service-learning. Your campus may be able to assign a work-study student to assist you, or you may be able to find a student who can do an independent study as your course assistant. Another option is to give one or two students in your class extra credit if they volunteer to be course assistants. However, using your own students as course assistants can be problematic if the students have to supervise or report on other students. Better tasks for such internal course assistants are helping to organize transportation, scheduling and materials preparation.

Campus Compact is a national organization dedicated to supporting service-learning work on college campuses. The annotated bibliography of this volume provides the web-

Table 6.2. Where to find people-power to help you run a service-learning course.

- The Service-Learning office
- VISTA volunteers trained and placed by Campus Compact
- Work-study students
- Students engaged in independent studies
- Students engaged in extra-credit projects, or an honors program
- Course assistants funded by service-learning grant funding

site information for national and state Campus Compact offices. The national Campus Compact office is located in Providence, Rhode Island, and 30 states have their own state Campus Compact offices. VISTA volunteers trained by Campus Compact are recent college graduates who are assigned for a year at a time to work on community-based projects. You can contact your state's Campus Compact to see if they run VISTA volunteer training and placement programs. There is a competitive application process to get a VISTA volunteer assigned to your campus. Your state's Campus Compact office can also provide you with information on sources of funding for service-learning endeavors, written materials on service-learning, and other forms of technical assistance.

Finding a Community Partner Organization: Using Mathematics for Public Problem Solving

In general, much of the service or volunteer work done by college students in their local communities is direct-service work, such as working with the elderly, with homeless families, with adult education students, or with children. However, many mathematics service projects are likely to fall into the realm of public problem solving—investigating with organizations how to do things better or more efficiently. Students may have a hard time recognizing this type of partnership as service work. In this case, it is especially important to help them see how the results of their work will "trickle down" and benefit individual members of the community.

On the other hand, some service-learning work in mathematics does involve working directly with a client base. For example, students can use their skills to help people understand personal finance, such as car loans vs. car leasing, credit card interest, and home mortgages. Faculty supervising these students need to emphasize the difference between community service and service-learning, in particular by emphasizing the importance of reflection and by linking the service experience with academic learning. The community partners also need to know that their partnership with service-learning students will be different than their partnerships with regular volunteers. Since the students are doing the service as part of an academic enterprise, they will come to the service with a different agenda and higher expectations about what they will learn from their experience.

In looking for a community partner, you need to decide how far and wide you want to go in your search. Take travel time for students into account, but don't turn down an exceptional project solely on the basis of significant travel. You may be fortunate enough to work at an institution that has a van or other transportation available to students through the service-learning program, community service program, or the student activities office. When campus vans or shuttles are not available, carpooling can work, but it requires advance planning and a back-up plan.

In working with local government, it is important to investigate what kind of infrastructure your city or town has. If you are interested in finding solutions for a complex problem, such as pollution or traffic, you may want to work with more than one agency or government office. Departments or agencies that lack the time or resources to share information with each other can benefit by working with a class of college students who can attack the problem from different angles by splitting into groups and working with several organizations or town offices. Networking is a good way to find a partner. A likely place to start is to identify organizations that are accustomed to collaborating with the university. The college offices mentioned in Table 6.1 can help you connect with such organizations.

The authors of chapters 2–5 offer many suggestions for sources of community partners in specific subject areas. Ideally, community partners for service-learning projects should be non-profit agencies, schools, or government agencies. Can you partner with a for-profit organization for a service-learning project? Providing a free service to a for-profit agency does not qualify as service just because it is free. The service rendered should respond to an unmet community need and contribute to the public good. If the for-profit business provides a valuable service to an underserved or socio-economically disadvantaged area, the project may qualify as public service. Helping a local business to comply with environmental standards might be a good service project conducted with a for-profit enterprise. For example, the Environmental Protection Agency has a long list of Voluntary Compliance Programs on their website at www.epa.gov/partners. Students could help a local government agency or business to participate in these programs, or they could help determine if the agency or business can afford to participate in the voluntary compliance program.

Faculty who are designing service projects for mathematics education courses should make sure that they consult with all of the right players in the school system when looking for a partner in the public schools. It may be easier to make an initial contact with a teacher or a parent group, but make sure that you also get principal or superintendent buy-in and approval before you establish your partnership.

Tips for Success in Working with a Community Partner

Once you have concluded that you have a good match for a project, it is time to determine whether your community partner will be a good organization to work with in a logistical sense. Table 6.3 lists a number of useful aspects to keep in mind at various stages of project planning. For example, make sure that the agency is organized and has enough staff to return phone calls to you and your students, produce necessary data or background infor-

Table 6.3. Tips for success in working with a community partner.

- Address a real community need
- Seek a community partner who has the time to communicate on a regular basis with you and your students
- Discuss the Academic Calendar with your community partner
- View your community partner as a co-instructor
- Educate your community partner about the difference between Service-Learning and Community Service
- Inform your community partner of your learning objectives for the project
- Set up a service-contract with your community partner and your students to clarify expectations
- Schedule regular check-in times for evaluation and feedback from your students and the community partner.

mation in a timely manner, and meet with students. There is a fine line between a non-profit agency that is in need of assistance because they are short-staffed, and a non-profit agency that is in need of assistance, but is too busy to make time to work with you. It may be difficult to determine the capacity of the organization ahead of time. Test the waters by checking with other academic departments who have worked with the organization. What was your experience contacting the organization? Did the organization return your phone calls in a timely manner? Is a staff member of the organization willing to put enough time into thinking carefully about what the organization's true needs are?

Be sure to spell out the academic calendar and your deadlines to the prospective partner organization. Many people outside of academia think of the academic calendar as ending in June, when in reality many students finish their classes the last week in April, and must complete their projects by that deadline. Academic institutions that operate on a quarter or a trimester system face other calendar challenges. Ask your partners about their schedules—do they have a large upcoming audit or project that will consume their time and make them unavailable to consult with your students? If so, the timing may not be good for collaboration. Discussing timelines before the semester begins will help define and shape the project. As you discuss the timeline, you may realize that you need to narrow the scope of your project, or rearrange your syllabus to make the project work. While planning your course, keep in mind that you may have to allow time to instruct students in certain mathematical concepts before the students can use these concepts to complete their service project.

When planning your interaction with the community partner, investigate ways that the community partner staff can help educate your students. Community partners have information about the community and the business world that can add value to your class. The partner can be invited to speak in class, or invited to suggest additional readings or other resources. The more the community partner knows about the philosophy behind service-learning, the more they will see themselves as teachers rather than just receivers of service. Thus, you should take time to educate your community partner about the difference between community service and service-learning and to inform your partner of the academic and civic learning objectives of your course.

As the course progresses, sharing excerpts of journal entries and student writings (with student permission) with the community partner is an effective way to help community partners understand the impact of the service-learning on students. However, be sure to read your students' writing before you decide to pass along their comments. When students encounter frustrations or setbacks during their work with community partners, they often express criticisms or comments about their community partners in their writings or class discussions. Usually, students' opinions change as their relationship continues with the organization, and they gain a better understanding of the reality of the work environment, or the lack of resources available to the organization. Even when the students' criticisms are valid, it may be best to keep such observations internal if you wish to continue to work with the community partner, while brainstorming ways that your organization can help improve their services. A discussion with your students about what kind of feedback would be productive and appropriate to share with the community partner can be an educational experience in itself.

A good way to clarify expectations on both sides is to set up a contract between your students and the agency about what services the students will perform, what the deliver-

Table 6.4. Points to include in student and community partner contract.

* Complete client and student contact information
* Days and hours of operation of the community partner organization
* Black-out dates when clients or students are unable to meet due to vacations, exams, or events in the workplace
* Agreed upon channels and means of communication
* Agreed upon turn-around time for answering emails or telephone calls
* Deadlines for stages of project
* Deadline for completion of project
* Final deliverables
* Plan for final presentation of deliverables to all stakeholders

ables will be, and what the agency will provide. Include as many dates as possible in the contract. It is important to clarify who at the agency will serve as a main contact for the students, who will supervise the students if necessary, and who will serve as a back-up contact person at the agency. Table 6.4 includes a list of points to include in the service-contract.

Lastly, in order for your students to get the most out of their service-learning projects, they should get off campus as much as possible. Community projects are much more meaningful and successful if the students spend time on site and experience the physical and socio-economic setting within which the community partner operates. Time spent consulting face to face with community partner staff is invaluable to the learning process. Such visits are time-consuming, but essential in order to develop a true partnership and for students to gain a deeper understanding of community dynamics. Observing and understanding how the organization works help students to produce results that are more useful for the organization.

Tips for Preparing and Working with Students on a Service-Learning Project

Some students arrive at college having participated in many hours of community service. Thus, it is important to explain to students how service-learning is different from community service and different from an internship. The goal of service-learning is to deepen academic learning through solving a public problem, including a healthy dose of reflection on the student's role in the process. To place your service-learning work in context, inform your class about other students and faculty who are engaged in service-learning at hundreds of respected academic institutions in the U.S. and worldwide, in almost all academic disciplines. Clearly explain the learning objectives of the service project and how they tie into the learning objectives of the course in general. Discuss your expectations for the relationship between the students and the community partner. Students should see the project as a consulting project and treat the people involved with the respect due to consulting clients. On the other hand, students should expect that they will be treated with

Table 6.5. Topics to discuss with students before they begin working with a community agency.

- The definition of service-learning
- Service-learning as an educational movement on a national/international level
- Course learning objectives: academic, interpersonal, civic
- Expectations for student/community partner relationships
- Assessment methods and grading of service-learning work
- Professional behavior (dress, writing style for email correspondence, etc.)
- Confidentiality, ethical behavior, liability
- Project logistics: schedules, transportation, communication

---

respect when interacting with the community partner. It is advisable to revisit the discussion of the rationale for the service-learning project and your course learning objectives at different points throughout the course, as service-learning provides students with experiences that are very different from those in their other academic classes.

Because many students will have previously participated in service in a non-graded, non-academic setting, they may be confused about the idea of being "graded" on their service project. Discuss up front your assessment methods for your course as a whole and particularly for the service-learning piece of your course. Explain in detail how your students will be graded on their service-learning work. Their grades should reflect the quality of the work that they produce and the level of understanding of mathematical concepts they can demonstrate, not just the number of hours they spend at a community agency.

To give your relationship with the community partner the best possible chance for success, discuss with the students the importance of professional behavior when they are introduced to the agency. Students should be reminded of good practices for writing professional emails and contacting people at the organization. For example, students are accustomed to using on-line instant messengers with their peers, and getting a response to emails within 24 hours. Students need to be aware that in the business or non-profit world, communication may not work that fast: people take sick days, have meetings first thing in the morning and may not check their email until late in the day. At some non-profit organizations, the email service may be unreliable. Creative solutions for making sure that the lines of communication stay open need to be found. If meetings are set up between students and an organization, students should be reminded that punctuality is essential, because people at the organization have budgeted valuable time to meet with them. It may be helpful to bring in a staff member from the career services office to lead a discussion with students about professional behavior and successful project management.

If students are assigned to work in groups on a service project, their collaboration provides another avenue for learning. It is a good idea to put in place mechanisms to monitor how the groups are functioning. Many faculty members have students submit periodic evaluations of themselves and their group members, so that faculty can determine if the students are putting in an equal amount of work and are working well together.

Do not underestimate the importance of clear instructions for students about transportation, meeting times, and communication. A course assistant can be very helpful with logistics—setting up meetings, arranging car-pooling, or photocopying materials that need to be shared. As a faculty member, your energy should be spent as much as possible on course design and learning goals, not on worrying about moving 20 of your students from Point A to Point B. The list of possible sources of "people-power" at your institution, given earlier in Table 6.2, may help you find substantial assistance with logistical tasks.

Besides transportation and scheduling, there are other logistical factors to investigate before beginning a project in the community. Check to see if the college needs the students to sign a liability release before working off campus. You may also consider having your students sign a confidentiality agreement, especially if the students are working with sensitive data or writing about the agency in journal entries. Students who work with children will probably have to fill out a criminal background check form to be processed by the state police. Criminal background check forms require students to provide their social security numbers, which newly arrived international students may not have. Sometimes students must also present proof that their vaccinations are up to date before they can work in the public schools. This information should be available from the college's student health center. The processing of forms can sometimes delay the start of a project, so research this aspect of the collaboration ahead of time.

Most public schools are accustomed to using volunteers from the community in the classroom, and should be able to inform you of the proper procedures for volunteer paperwork and how to expedite the process. Elementary school tutoring programs have found it useful to provide their student volunteers with large nametags identifying which institution they come from, and which program they are working on. The nametags cut down on the number of times that college students are stopped in the hallways and asked "who are you and why are you here?" Ask the principal to inform the entire school in advance that your students will be working in the building, and to encourage all the teachers to be welcoming. College students are often nervous during their first visits to the schools, and the more smiling faces they encounter the better. Clarify appropriate dress codes for students working in the public schools, as well as any special school rules (sign-in procedures, no gum, no giving candy to children, etc.). For some tutoring programs, schools may also need to seek parental permission for their students to participate in the program. Time should be allowed for the process of distributing and collecting permission slips from children.

Regardless of what kind of organization your students are working with, it is a good idea to ask your contact to introduce your students to as many people in the organization as possible. The staff will then know why the students are there, and be more welcoming as a consequence. On the other hand, students should be encouraged to take notes and remember the names and job titles of important staff in the organization, especially those who will serve as resources throughout the project.

Timeline for Designing and Teaching a Service-Learning Course

Just as text and teaching materials are chosen well ahead of the start of the semester, the service-component of a class generally should be worked out several months in advance of the first class meeting. The instructor and the students have a better experience when the instructor treats the service-learning experience as an integral part of the course, rather than as a piece that is tacked on. Table 6.6 provides a suggested timeline and checklist for planning a service-learning course.

When creating the service-learning assignment for your students, set up a timeline for different stages of the project, rather than establishing only one final deadline for the project's completion. Working with the community partner can take more time and person-to-person coordination than normal classwork, and thus requires the ongoing participation of the students and the community partners. If students must submit progress reports on a regular basis, they learn to budget their time more wisely and to meet with their community partners on a regular basis.

In order to deepen the learning from a service-learning experience, it is imperative to include regular writing, discussion, or presentation assignments that require students to think critically about their service experience, and to take the time to examine the connec-

Table 6.6. Timeline and Checklist for a Service-Learning Course.

Several months before semester begins	Identify mathematics or quantitative learning objectives for a course Identify other learning objectives for course in the areas of civic engagement and communication or interpersonal skills Identify a real-world application for the mathematical content of your course Investigate Community Partner sites Outline project with Community Partner
Two weeks before semester begins	Check in with Community Partner to make sure that project will move forward as planned earlier
Before 1st day of class	Develop course syllabus and project timeline
1st or 2nd day of class	Present service-component of course to students
1st two weeks of class	Assign students to groups to work on project Finalize logistics for transportation, etc. Before students meet with community partners for the first time, conduct pre-service reflection/critical thinking exercise
Weekly or Biweekly during semester	Conduct reflection exercises with students and collect progress reports
Several times during semester	Check in with community partner, by phone or in person
Mid-semester	Set dates for final presentation to community partner
End of semester	Schedule feedback and evaluation for project with all stake-holders Share your results with other interested parties at your college and in the community Revise project and course for next semester based on student and community feedback

tions between their service experience and the other academic material presented in the course. Many instructors make pre-service, mid-semester, and end of semester reflection assignments. Reflection can take many forms, including writing, whole class discussions, small group discussions, and presentations. See Table 6.7 for sample pre-service and middle or end of semester reflection and critical thinking assignments to accompany community work.

Designing and leading a service-learning course is hard work, and it is easy to run out of time. Make sure you set aside time to conduct an evaluation of how students and community partners have experienced the project. Determine in advance how you will measure your students' learning. In your evaluation instrument, you also can pose questions about your students' attitudes toward community work, corporate responsibility, and civic engagement. Many evaluation tools to measure civic engagement have been created by faculty working in service-learning and are available for your use. It is often helpful to conduct pre- and post-service surveys if you are interested in measuring attitudes towards community engagement. (See the annotated bibliography at the end of this chapter.) Community partners should also evaluate the project, providing information on the quality of their interaction with university staff and students, and on the extent to which their needs were met, with suggestions for improving the collaboration.

Organize an official closing presentation for the project that includes all of the stakeholders. This can be a formal gathering where your students present their findings to the community partner. You may wish to invite other faculty or college officials, and ask the community partner whom they would like to include. A non-profit agency may want to invite its board members. The presentation can be on campus or in the community. Faculty members have found that students create higher quality presentations when they know that they will be presenting to an audience that includes working professionals and community members in addition to faculty and fellow students. The final presentation is also a good time to thank publicly your community partners for the time and effort they have spent working with your students. Equally important, the community partner can thank you and your students for your valuable service. When you have finished your project in the community, why not publicize your efforts? Tell your media relations office about what you are doing. As the news spreads, other sites may seek your help and more projects will be generated for the college.

In follow-up conversation with your community partner, check that the organization received all of the reports and data that were produced by your students. Does the organization have the resources to actually do something with the report that you gave them? You may determine that it makes sense for the organization to work with you or your academic department on a regular basis, or that another department in the university can pick up where your class left off. It's also helpful to decide if the organization is a good site for future interns from your institution, especially if they are in need of assistance from students who can devote more hours to specific projects.

Requiring Service-Learning in Your Course

When faculty members decide that they would like to include a service-learning component in a course, they often face the question of whether or not they will require all of their

Table 6.7. Sample questions for reflection/critical thinking exercises, designed to harvest learning from service experiences.

For Pre-Service Reflection: (before 1st visit to the site)

1. What mathematical skills will you need to use in order to complete your project?
2. What other skills will you need to use in order to successfully complete your project? (technology skills, communication skills, problem-solving skills, writing skills, etc.)
3. What aspects of this project do you expect to be exciting, challenging, frustrating, rewarding for you?
4. What are your expectations about the site where you will do your project—what will it look like? What will the people be like at the site?
5. How does your project relate to your mathematics class as a whole?
6. What kind of help will you need from your instructor and from the service-learning center to make your project successful?
7. What assumptions might the staff at the community partner organization have about college students engaged in service-learning? How can you make the best possible impression at the community partner organization?

For Mid-Semester or End-of-Semester Reflection Sessions

Questions about Student Skills:

1. What mathematical skills did you use in order to successfully complete your service project?
2. What people skills and communication skills did you use to successfully complete this project?

Questions about Success of the Project:

3. What was the best part of your service-learning experience?
4. What would you change about the design of the service-learning project?
5. What specific comments and recommendations do you have regarding communicating with the community partner, working in groups, logistics, and the time-commitment required for the project?

Questions about the Community Partner:

6. What limitations or obstacles does your Community Partner face? (lack of money, space, staff, etc.) How do service-learning students help the organization? Based on your observations, what further technical assistance does your agency need? Where could your organization turn to in order to get that assistance?
7. What societal problems is your agency addressing? How do they address them?

Questions about Service after Graduation:

8. Do you envision yourself lending your mathematical skills or time to community organizations or individuals in the future? If yes, what kind of work would you find satisfying? Direct service work? Serving on a non-profit agency Board of Directors?
9. How would you define "corporate responsibility to the community"? Do you believe that large corporations have a responsibility to help out the community that they operate in? Do you believe that colleges and universities have a responsibility to assist the community that they reside in?

students to participate in the service-learning project. The various project examples in this book illustrate both approaches. One way of framing this issue is to consider that many courses require students to participate in group projects or field visits that require substantial out-of-class time, and service activities involve a comparable type of commitment. The important difference between service-learning projects and student group projects is that there is less flexibility in scheduling time spent on the service project. When students are required to work on a group project, they can meet at their convenience. By contrast, a service-learning project most likely will require students to meet with an outside organization during business hours, or to work with children during public school hours. Such time constraints can be problematic for students who have commitments to a job, to family, or to athletics. If you require students to participate in service-learning, it is best to indicate in the course listing that service-learning is required in your course. You should also discuss the service-learning project with your students on the first day of class and before the add-drop deadline at your institution. To require service-learning you may need to obtain permission from your department head or other university official, and the issue is more delicate if your course is a required course rather than an elective.

Nevertheless, many arguments can be made for the value of involving all of your students in service-learning. In general, faculty members have been most satisfied with the learning outcomes of service-learning projects when the service work is closely integrated into the material of the course. If all students participate in the project, it is easier to justify devoting class time to making the connection between the service activity and the academic content of the class on a regular basis. Requiring all students in the class to participate in the project also emphasizes to the students that the project is an integral and important part of the course, and is directly related to academic learning, rather than being just a "feel good" activity.

If a faculty member wants to engage in service-learning, but is not ready or able to require service-learning as a part of the course, there are other options for giving students academic credit for their service work. Some institutions offer the option to students to register for the service-learning activity as a separate one-credit class that is "attached" to another academic course. This type of program is called the "4th credit option" at Bentley College in Waltham, MA and is referred to as the "Plus-one" program at Rivier College in Nashua, New Hampshire. Some faculty members have offered service-projects as an extra-credit assignment. Others offer the students a choice between a research project and a service-learning project. If you do make the service-learning project optional, it is important to plan ahead of time to meet with service learners outside of class to monitor their progress and to conduct reflection activities with them. Ideally, you can build into your class some ways for the service-learning students to share their learning experiences with the rest of the class.

One danger in giving students the option of participating in the service-learning project is that students may want to withdraw from the project after making an initial commitment. This can cause problems if you have already worked out the project with the community partner and you are counting on the participation of a certain number of students. If the service project is optional, make it clear that once students commit to the project in the community, they must stay with it for the duration of the semester. Finally, if you offer the service-learning project as an option, you will not know until the semester begins how many students will participate in the project. This poses some challenges when planning the service-learning project with your community partner.

Levels of Logistical Complexity of Service-Learning Projects

Does incorporating service-learning into a mathematics or other quantitative course seems like a daunting task? Keep in mind that service-learning projects can be done successfully at various levels of complexity. Logistical complexity refers to the amount of effort that needs to be spent on setting up and planning a project, not to the level of mathematical complexity involved in the projects. The same mathematical concepts can be incorporated into service-learning projects that are fairly easy to arrange, or into service-learning projects that require more time and effort to plan and to execute. For example, mathematics education students can be involved in tutoring in elementary school classrooms. Since the teachers provide the lesson plans and the materials, this project involves less preparation time. Having college students come up with their own lesson plans adds to the time and effort required for the project, but the learning outcomes will probably be greater for the college students because they take a more active role in the project. Arranging a free-standing mathematics program for children such as a Family Math Night, or an ongoing America Counts tutoring program, requires even more effort in the logistics and planning. Faculty engaging in service-learning for the first time may want to start with projects that require less logistical planning, and adjust or expand the service-learning projects during subsequent offerings of the course.

Incorporating service-learning into a course does require a time commitment from faculty, but the potential rewards of a service-learning course are substantial. Many instructors report that students experience increased learning and retention of mathematical concepts as a result of service-learning activities. Courses that include successful service components result in increased interest and dedication from students. In working on service-learning problems, students can get a glimpse of how mathematics is relevant in real-life situations. A service project may allow students to see how they can turn the study of mathematics into a career in industry or the public sector. By collaborating with their peers and with community partners, students are given the opportunity to develop interpersonal and professional skills, and to function as engaged citizens.

Annotated Bibliography

Print Resources to Aid in Service-Learning Course Construction

Service-Learning Course Design Workbook, Michigan Journal of Community Service-learning. Order from website: www.umich.edu/~mjcsl/

> This concise volume provides an explanation of basic service-learning concepts and best practices. More importantly, it provides faculty with charts and worksheets to help identify service-learning course learning objectives, learning strategies, and assessment methods in two areas: enhanced academic learning and purposeful civic learning. An extremely practical publication.

Battistoni, Richard. *Civic Engagement Across the Curriculum. A Resource Book for Service-Learning Faculty in All Disciplines*. Campus Compact, Providence, RI, 2001. Order from Campus

Compact website: www.compact.org

> This book provides a deep exploration of the role of and the importance of Civic Engagement in service-learning endeavors. In addition to the thoughtful chapters on such topics as "What is Good Citizenship? Conceptual Contributions from Other Disciplines" and "Being Attentive to the Civic Dimensions of Service-Learning" it includes several useful Appendices such as "Reflection Questions that Tap Civic Dimensions," "Public Problem Statement and Organizational Issue Research," and many sample assignments.

Heffernan, K. *Fundamentals of Service-Learning Course Construction*, Campus Compact, Providence, RI, 2001. Order from Campus Compact website: www.compact.org

> This book provides faculty with a plan for designing a service-learning course. It describes several models of service-learning courses, and discusses in details elements such as designing course objectives and rationales, describing the service-assignment and grading. It provides many sample assignments and syllabi. The volume does not include sample mathematics course syllabi, but mathematics faculty can still find this publication useful in their own course planning.

General Background on Service-Learning

Introduction to Service-Learning Toolkit. Campus Compact, Providence, R.I., 2003. Order from Campus Compact website: www.compact.org

> This book is a collection of articles that discuss many topics related to service-learning such as the definition of service-learning, theory, pedagogy, community partnerships, reflection, assessment, redesigning curriculum, etc.

Eyler, Janet, Giles, Dwight E., Schmiede, Angela. *A Practitioner's Guide to Reflection in Service-Learning: Student Voices and Reflections*, Nashville, TN, Vanderbilt University, 1996.

> This book is a very user-friendly guide to reflection. It includes a description of many types of reflection and description of reflection activities and exercises. It also examines how reflection helps students in the learning process. It can be ordered from the National Service-Learning Clearinghouse.

Essential Service-Learning Resources, Campus Compact, Providence, RI, 2003.
Order from Campus Compact website: www.compact.org

> This booklet is available free of charge from Campus Compact website. It includes an extensive list of service-learning resources available in print as well as statistics on service-learning and answers to many frequently asked questions about service-learning.

Michigan Journal of Community Service-Learning. See webpage for ordering information: www.umich.edu/~mjcsl/

> The only scholarly journal devoted entirely to service-learning. The Michigan Journal of Community Service-learning is a national, peer-reviewed journal consisting of articles written by faculty and service-learning educators on research, theory, pedagogy, and issues pertinent to the service-learning community.

Assessment

Eyler, Janet, and Giles, Dwight E., Jr., *Where's the Learning in Service-learning?* Jossey-Bass, San Francisco, CA, 1999.

> This book provides the history and theoretical basis for service-learning, and presents best practices for teaching service-learning courses based upon extensive research. Included in the book are evaluation tools that measure students' attitudes toward civic engagement and other learning outcomes.

Driscoll, Amy, Gelmon, Sherril, Holland, Barbara, Kerrigan, Seanna, Longley, M.J., Spring, Amy. *Assessing the Impact of Service-learning: A Workbook of Strategies and Methods,* Portland State University, Portland, Oregon, 1997.

> Provides strategies and methods for assessing and measuring the impact of service-learning on students, faculty, community, and institutions; includes case studies of service-learning courses.

On-line Resources

National Service-Learning Clearinghouse Website: www.servicelearning.org

> The National Service-Learning Clearinghouse is funded by the Corporation for National and Community Service. Under the link to the NSLC library, there is a link to concise fact sheets on evaluation tools, funding sources, community partnerships, and many other useful subjects related to service-learning.

Campus Compact: www.compact.org

> Campus Compact is a national coalition of more than 860 college and university presidents committed to the civic purposes of higher education. This site has links to information on funding opportunities, recent publications and effective practices in Service-learning.

State Campus Compacts: www.compact.org/state

> This link from the national Campus Compact site provides contact information and links to Campus Compact organizations by state. State Campus Compacts provide information on funding opportunities and other resources. They also organize conferences and workshops for faculty working in service-learning.

Community College National Center for Community Engagement: www.mc.maricopa.edu/other/engagement/

> This site provides information on service-learning programs at community colleges. The site includes service-learning tools and resources as well as information on community-college specific conferences, publications, and service-learning awards.

American Association for Higher Education: www.aahe.org/service/

> This site provides information on AAHE's series on Service-Learning in the Disciplines, the Campus Compact Consulting Corps, and links to general information on service-learning and models of good practice for service-learning programs.

Service-Learning Website and List-Serve hosted by University of Colorado at Boulder: csf.colorado.edu/sl/

> Includes a link to join an on-line service-learning discussion group, a link to discussion group archives, and a Guide to College and University Service-Learning Programs, Courses and Syllabi.

America Counts website: www.ed.gov/americacounts/

> This site provides information on the America Counts tutoring program for elementary school children. The site includes information on the history of the initiative, curriculum, and how to organize such a program.

An Inventory of Your Service-Learning Partnership, Campus Community Partnerships for Health, 1999.

> Can be found on the Campus Community Partnerships for Health's website: www.ccph.info/ —under the tools and resources menu in the service-learning section

This inventory helps faculty to evaluate their present community partnership and to decide how to improve the partnership. Ideal for faculty who have already developed a relationship with a community partner.

Sample Service-Learning Agreement, on Campus-Community Partnerships for Health webpage: www.ccph.info/ —under the tools and resources menu in the service-learning section

Provides an example of a contract between service-learning students and a community partner agency, signed before the commencement of a service-learning project.

Syllabus Revision Procedures, Dr. Edward Zlotkowski, on Campus Community Partnerships for Health webpage: www.ccph.info/ —under the tools and resources menu in the service-learning section.

Provides a helpful one-page summary of tips for revising a syllabus in order to integrate a service-learning project into a course.

About the Authors

Jennifer Webster, MA, EdM, coordinates tutors and resources at the Harvard Bridge to Learning and Literacy Program. She was formerly the Coordinator of Academic Programs at the Bentley College Service-Learning Center in Waltham, Massachusetts, where she served as principal liaison between faculty, students, and community partners engaged in service-learning projects. She has worked closely with many organizations in the community and with faculty and students from all academic disciplines.

Charles Vinsonhaler, PhD, is a Professor of Mathematics at the University of Connecticut in Storrs, Connecticut, where he recently completed a six-year term as department head. He conducts research in algebra (abelian groups, rings) and has side interests in actuarial science, problem solving, and new pedagogies such as service-learning.

JSC Willey Library
337 College Hill
Johnson, VT 05656